SOURCEBOOK OF ADVANCED POLYMER LABORATORY PREPARATIONS

SOURCEBOOK OF ADVANCED POLYMER LABORATORY PREPARATIONS

Stanley R. Sandler
Elf Atochem North America
King of Prussia, Pennsylvania

Wolf Karo
Consultant to the Chemical Industry
Huntingdon Valley, Pennsylvania

ACADEMIC PRESS

San Diego   London   Boston   New York   Sydney   Tokyo   Toronto

This book is a guide to provide general information concerning its subject matter: it is not a procedural manual. Synthesis of chemicals is a rapidly changing field. The reader should consult current procedural manuals for state-of the-art instructions and applicable government safety regulations. The Publisher and the authors do not accept responsibility for any misuse of this book, including its use as a procedural manual or as a source of specific instructions. The users of this book knowingly and voluntarily assume all the risks of any and all injuries that may result from performing any experiment described herein.

This book is printed on acid-free paper. ∞

Copyright © 1998 by ACADEMIC PRESS

All Rights Reserved.
No part of this publication may be reproduced or transmitted in any form or by any means, electronic or mechanical, including photocopy, recording, or any information storage and retrieval system, without permission in writing from the publisher.

Academic Press
a division of Harcourt Brace & Company
525 B Street, Suite 1900, San Diego, California 92101-4495, USA
http://www.apnet.com

Academic Press Limited
24-28 Oval Road, London NW1 7DX, UK
http://www.hbuk.co.uk/ap/

Library of Congress Card Catalog Number: 98-84428

International Standard Book Number: 0-12-618605-7

PRINTED IN THE UNITED STATES OF AMERICA
98 99 00 01 02 03 EB 9 8 7 6 5 4 3 2 1

CONTENTS

Preface xv
Introduction xvii

1 POLYMERIZATION OF STYRENE AND OTHER OLEFINS

1. BULK POLYMERIZATION 3
   1-1. Preparation of Polystyrene by the Bulk Thermal Polymerization of Styrene 4
2. SUSPENSION POLYMERIZATION 5
   1-2. Preparation of Polystyrene by Suspension Polymerization 6
3. EMULSION POLYMERIZATION 6
   1-3. Emulsion Polymerization of Styrene 7
4. ANIONIC POLYMERIZATIONS 8
   1-4. Preparation of Polystyrene by the Polymerization of Styrene Using Sodium Naphthalene Catalyst 11
5. PREPARATION OF MONODISPERSED MICROSPHERES OF POLYSTYRENE 12
   1-5. Monodispersed Crosslinked Microspheres by Dispersion Polymerization 13
6. POLYMERIZATION OF OTHER OLEFINS USING A CATIONIC POLYMERIZATION PROCESS 14
   1-6. Preparation of Butyl Rubber (Copolymerization of Isobutylene with Isoprene Using BF_3 Catalyst) 16
7. POLYMERIZATION OF OTHER OLEFINS BY AN ANIONIC POLYMERIZATION PROCESS 17
   1-7. Preparation of Polybutadiene by the Amylsodium-Catalyzed Polymerization of Butadiene 17

8. CATALYTIC BREAKTHROUGHS IN POLYMERIZING OLEFINS 17
References 19

2 POLYESTERS

2-1. Preparation of Poly(ethylene terephthalate) 27
References 29

3 POLYCARBONATES

3-1. Preparation of Poly(2,2-dimethyl-1,3-propanediol carbonate) [poly(neopentylene carbonate)] 32
References 33

4 POLYAMIDES

4-1. Preparation of Poly(hexamethylenesebacamide) (Nylon 6-10) by the Interfacial Polymerization Technique 38
References 38

5 POLYMERIZATION OF ALDEHYDES

5-1. Preparation of Polyacetaldehyde by Cationic Polymerization of Acetaldehyde Using BF_3 Catalysis 42
5-2. Polymerization of Acetaldehyde Using Potassium Triphenylmethoxide Catalysis 43
References 44

6 POLYMERIZATION OF EPOXIDES AND CYCLIC ETHERS

6-1. Preparation of Polyoxypropylene Glycol by the Room-Temperature Polymerization of Propylene Oxide Using KOH 47
6-2. Polymerization of *l*-Propylene Oxide with Solid KOH Catalyst 49
6-3. Polymerization of Ethylene Oxide Using $SnCl_4$ (or BF_3) Catalyst 49
References 52

7 POLYUREAS

1. REACTION OF DIAMINES WITH DIISOCYANATES 54
7-1. Preparation of a Polyurea from Bis(3-aminopropyl) Ether and 1,6-Hexamethylene Diisocyanate 54

2. REACTION OF DIAMINES WITH UREA 55
   7-2. Preparation of Poly(4-oxyhexamethyleneurea) 55
   References 56

8 POLYURETHANES

   8-1. Preparation of a Polyurethane Prepolymer 61
   8-2. Preparation of Poly[ethylene methylene bis(4-phenylcarbamate)] 62
   References 62

9 THERMALLY STABLE POLYMERS

   9-1. Preparation of Poly(2,6-dimethyl-1,4-phenylene oxide) by Polymerization of Silver 4-Bromo-2,6-dimethylphenolate 66
   9-2. Preparation of Poly(2,6-dimethyl-1,4-phenylene oxide) Using $CuCl-Pyridine-O_2$ 68
   References 69

10 POLYMERIZATION OF ACRYLIC MONOMERS

1. POLYMERIZATION OF ACRYLIC ACIDS AND RELATED COMPOUNDS 73

   10-1. Preparation of Isotactic Poly(acrylic acid) 73
   10-2. Preparation of Poly(methacrylic acid) by Suspension Polymerization with a Monomer-Soluble Initiator 75
   10-3. Polymerization of Methacrylic Acid in an Aqueous Solution 76
   10-4. Charge-Transfer Complex Polymerization of Acrylic Acid-Acrylamide 77
   10-5. Bottle Polymerization of Itaconic Acid 79

2. POLYMERIZATION OF ACRYLAMIDE AND RELATED AMIDES 81

   10-6. Suspension Copolymerization of Acrylamide and Acrylic Acid 85
   10-7. Polymerization of Acrylamide in Aqueous Solution 86
   10-8. Preparation of Poly(acrylamide) Gels 86
   10-9. Polymerization of Methacrylamide in Toluene 87
   10-10. Emulsion Copolymerization of Styrene and Acrylamide 89
   10-11. Inverse Emulsion Polymerization of Aqueous Acrylamide in Toluene 90

10-12. Anionic Polymerization of N,N-Dimethylacrylamide 91
10-13. Hydrogen-Transfer Polymerization of Acrylamide in Pyridine 93
3. POLYMERIZATION OF ACRYLONITRILE AND RELATED MONOMERS 94
10-14. Slurry Homopolymerization of Acrylonitrile 97
10-15. Solution Polymerization of Acrylonitrile in Zinc Chloride Solution 99
10-16. Emulsion Polymerization of Acrylonitrile 100
10-17. Emulsion Polymerization of Methacrylonitrile 101
10-18. Monosodium Benzophenone-Initiated Polymerization of Acrylonitrile 102
4. POLYMERIZATION OF ACRYLIC AND METHACRYLIC ESTERS 103
10-19. Generalized Procedure for the Preparation of Polymer Sheets 108
10-20. Suspension Polymerization of Methyl Methacrylate 110
10-21. Solution Copolymerization of Glycidyl Methacrylate and Styrene 111
10-22. Preparation of a Crosslinked Polymer Gel 112
10-23. Emulsion Polymerization of Ethyl Acrylate (Thermal Initiation) 114
10-24. Redox Emulsion Polymerization of Ethyl Acrylates 115
10-25. Preparation of Isotactic Poly(methyl methacrylate) 116
10-26. Group Transfer Polymerization of Methyl Methacrylate 118
References 119

11 AMINO RESINS

1. UREA-FORMALDEHYDE CONDENSATIONS 126
11-1. Preparation of Urea-Formaldehyde Textile Resins (F : U Ratio 2 : 1) 126
2. MELAMINE-ALDEHYDE CONDENSATIONS 127
11-2. Preparation of Hexamethylolmelamine Prepolymer 128
References 129

12 PHENOL-ALDEHYDE CONDENSATIONS

12-1. Preparation of a Phenol-Formaldehyde Resole 131
12-2. Preparation of a Phenol-Formaldehyde Casting Resin 132
References 134

13 EPOXY RESINS

1. ANALYSIS OF EPOXY RESINS 136
2. CURING-POLYMERIZATION REACTIONS OF EPOXY COMPOUNDS AND RESINS 137
   13-1. Curing Reaction of Bisphenol A Diglycidyl Ether with Poly(adipic acid anhydride) 137
References 139

14 ALKYD RESINS

14-1. Preparation of Poly(glyceryl phthalate) Alkyd Resin 144
References 147

15 POLYACETALS AND POLY(VINYL ACETALS)

1. POLYACETALS FROM FORMALDEHYDE 151
   15-1. Preparation of the Polyformal of Diethylene Glycol 151
2. POLYACETALS FROM SUBSTITUTED ALDEHYDES 151
   15-2. Preparation of a Polyacetal by the Reaction of Benzaldehyde with Bis(2-hydroxyethyl) Sulfide 152
   15-3. Preparation of a Polyacetal by the Reaction of Pentaerythritol with a Mixture of 1,3- and 1,4- Cyclohexane Dialdehydes 153
3. POLY(VINYL ACETALS) 153
   15-4. General Procedure for the Preparation of Poly(vinyl acetals) 155
   15-5. Preparation of Poly(vinyl butyral) 155
References 156

16 POLY(VINYL ETHERS)

1. CATIONIC POLYMERIZATION 159
   16-1. Preparation of a Poly(vinyl *n*-butyl ether) Using Stannous Chloride Catalyst 160
   16-2. Preparation of Poly(vinyl *n*-butyl ether) Using Boron Trifluoride Etherate Catalyst 160

2. FREE-RADICAL POLYMERIZATION   161
   16-3. Preparation of Poly(vinyl ethyl ether) Using Di-*tert*-butyl Peroxide Initiator   162
   16-4. Copolymerization of Maleic Anhydride with Ethyl Vinyl Ether   162
References   163

17 POLY(*N*-VINYLPYRROLIDONE)

1. BULK POLYMERIZATION PROCEDURES   167
   17-1. Bulk Polymerization with AIBN   167
   17-2. Bulk Polymerization of a PVP-Based Hydrogel   169
2. AQUEOUS SOLUTION PROCEDURES   170
   17-3. Generalized Procedure for Polymerization with Ammonia-Hydrogen Peroxide   170
3. CATIONIC POLYMERIZATION   172
   17-4. Copolymerization *N*-Vinylpyrrolidone and Methyl Methacrylate with Triethylboron   172
4. COMPLEX FORMATION   173
References   174

18 SILICONE RESINS (POLYORGANOSILOXANES) OR SILICONES

   18-1. Silicone Resins from Phenyltrichlorosilane   177
   18-2. Silicone Resins by Cyclohydrolysis of Dimethyldichlorosilane and Methyltrichlorosilane   178
References   178

19 OLEFIN-SULFUR DIOXIDE COPOLYMERS

   19-1. Preparation of Octene-Sulfur Dioxide Copolymers   181
   19-2. Preparation of Propylene-Sulfur Dioxide Copolymers   182
References   183

20 SULFIDE POLYMERS

1. CONDENSATION REACTIONS   185
2. OXIDATION REACTIONS   187
   20-1. Preparation of a Polydisulfide by the Oxidation of Hexamethylenedithiol with Bromine   187

3. POLY(ARYLENE SULFIDES)   188
  20-2. Preparation of Poly(phenylene sulfide) by the Reaction of p-Dichlorobenzene and Sodium Sulfide in N-Methyl-2-pyrrolidone   189
References   192

21 POLYMERIZATION OF MONO- AND DIISOCYANATES

1. HOMOPOLYMERIZATION OF MONOISOCYANATES TO 1-NYLON   194
  21-1. General Method for the Preparation of 1-Nylon   195
2. POLYMERIZATION OF DIISOCYANATES TO GIVE POLYCARBODIIMIDES   198
  21-2. Polymerization of 2,4-Toluene Diisocyanate   199
3. POLYMERIZATION OF POLYISOCYANATES TO POLYISOCYANURATES   201
References   201

22 POLYOXYALKYLATION OF HYDROXY COMPOUNDS

1. CHEMISTRY   204
2. POLYOXYALKYLATION OF ALCOHOLS AND DIOLS   207
  22-1. Preparation of Ethyleneglycol Monoethyl Ether   208
  22-2. Polyoxypropylation and Polyoxyethylation of 2,6,8-Trimethyl-8-nonanol   208
  22-3. Preparation of Polyoxypropylated Stearyl Alcohol   209
  22-4. Preparation of Polyoxypropylene Glycol   209
3. POLYOXYALKYLATION OF POLYHYDROXY COMPOUNDS   210
  22-5. Polyoxypropylation of Sucrose   210
  22-6. Polyoxypropylation of Sorbitol   211
  22-7. Preparation of Polyoxyethylated Pentaerythritol   212
4. POLYOXYALKYLATION OF PHENOLS   212
  22-8. Preparation of Polyoxyethylated Nonylphenol   213
  22-9. Polyoxypropylation of p-tert-Butylphenol   213
References   214

23 POLYMERIZATION OF VINYL ESTERS

1. MONOMER PURIFICATION 219
2. INITIATION, INHIBITION, AND RETARDATION OF POLYMERIZATION 222
3. POLYMER STRUCTURE 223
4. REACTIVITY RATIOS OF VINYL ESTERS 227
5. CONFORMATION OF POLY(VINYL ESTERS) 229
6. BULK POLYMERIZATION 230
   23-1. Sealed-Tube, Bulk Polymerization of Vinyl Acetate 232
7. SOLUTION POLYMERIZATION 232
   23-2. Polymerization of Vinyl Acetate in *tert*-Butyl Alcohol 233
8. SUSPENSION POLYMERIZATION 234
   23-3. Suspension Polymerization of Vinyl Acetate (Control of Molecular Weight by Variation in Initiator Level) 236
9. EMULSION POLYMERIZATION 237
   23-4. Emulsion Polymerization of Vinyl Acetate-Potassium Persulfate-Sodium Lauryl Sulfate System 239
   23-5. Emulsion Ter-polymerization of Vinyl Acetate, Butyl Acrylate, and Vinyl Neodecanoate (Seeded Process with Gradual Monomer and Initiator Additions) 242
10. NONAQUEOUS DISPERSION POLYMERIZATION 243
    23-6. Preparation of Poly(vinyl acetate) in Nonaqueous Dispersion 243
11. RADIATION-INITIATED POLYMERIZATION 244
References 245

24 POLYMERIZATION OF ALLYL ESTERS

1. POLYMERIZATION OF ALLYL ACRYLATE AND METHACRYLATE 252
   24-1. Solution Polymerization of Allyl Methacrylate 253
2. POLYMERIZATION OF DIALLYL CARBONATES 253
   24-2. Solution Polymerization of Diethylene Glycol Bis(allyl carbonate) 254
3. POLYMERIZATION OF DIALLYL ESTERS OF PHTHALIC ACIDS 255
   24-3. Bulk Polymerization of Diallyl *o*-Phthalate with High-Temperature Initiator 256

24-4. "Suspension Polymerization" of Diallyl *o*-Phthalate   257
References   259

25 POLY(VINYL ALCOHOL)

25-1. Preparation of Poly(vinyl alcohol) by the Alcoholysis of Poly(vinyl acetate)   263
References   264

26 INTRODUCTORY NOTES ON EMULSION POLYMERIZATION TECHNIQUES

1. INTRODUCTION   266
2. DEVELOPING AN EMULSION POLYMERIZATION RECIPE   268

26-1. Emulsion Polymerization of Styrene   273
26-2. Preparation of Butadiene-Styrene Copolymers by the Emulsion Polymerization Technique   273
26-3. Preparation of Acrylonitrile-Butadiene Copolymers by Emulsion Polymerization   274
26-4. Emulsion Copolymerization of Styrene and Acrylamide   274
26-5. Emulsion Polymerization of Vinyl Acetate without Surfactant   275
26-6. Emulsion Ter-polymerization of Vinyl Acetate, Butyl Acrylate, and Vinyl Neodecanoate (Seeded Process with Gradual Monomer and Initiator Additions)   275
26-7. Emulsion Polymerization of Ethyl Acrylate (Thermal Initiation)   276
References   277

Subject Index   279

PREFACE

The purpose of this text is to provide a ready source of reliable procedures for those involved in polymer syntheses in either academic or industrial laboratories. It will be useful to instructors and students seeking to find additional laboratory preparations or as a source of models to guide related polymer preparations. For example, the section on emulsion polymerization outlines the techniques used to prepare this class of polymers. Where possible, polymers are grouped according to their family, such as styrene-based polymers or acrylic monomers. Industrial chemists will also appreciate the tables of data and references to the literature, which include numerous patents.

The material used in this sourcebook is taken for the most part from chapters in our three-volume set of *Polymer Syntheses*, second edition. Additional material was used where needed, and precautions and notes are shown to help in carrying out the preparations.

We extend our appreciation to our families for their encouragement in this project and to the Academic Press staff for their efforts in guiding this work through the publication process.

Stanley R. Sandler

Wolf Karo

INTRODUCTION

The *Sourcebook of Advanced Polymer Laboratory Preparations* has been designed as a convenient source for synthetic procedures of wide utility for both students and industrial chemists. Students can use the *Sourcebook* to find additional experiments to illustrate concepts in polymer chemistry and to augment existing laboratory texts or manuals. The industrial chemist, who is frequently required to prepare a variety of polymer preparations, will find this bench-top manual a useful guide to the procedure or technique to use. In addition, the literature references direct the reader back to the original publications or patents describing the procedure.

Each section or chapter describes the synthesis of a given class of polymers and gives a brief introduction that summarizes the major procedures. One or more representative preparations are given, but the reader can obtain more information from the authors' original books in the three-volume set of *Polymer Syntheses*.

One unique feature of the *Sourcebook* is that it provides details on the emulsion polymerization techniques from one of the authors (W.K.), who has had many years of experience in this field.

The *Sourcebook* assumes that the student has already taken organic chemistry and has experience in the laboratory with the typical glassware normally used in these preparations.

1. SCALE OF OPERATIONS AND MONITORING OF REACTIONS

Most preparations cited can be scaled *down* provided that microware is available. The advantage is that less waste disposal is required.

Preparations should *not* be scaled up unless this is done *gradually* to determine whether the exothermic reaction can be safely tolerated by the equipment being used. All glassware should be free of cracks.

The reactions in most cases can be easily monitored with gas chromatography, infrared spectroscopy, ultraviolet spectroscopy, and thin-layer chromatography. Where available, a nuclear magnetic resonance (NMR) instrument can also be used very effectively to follow the course of the reaction and to determine the structures of the products.

A. SAFETY

All experiments should be carried out in a good fume hood with personal protection including safety eye glasses, a laboratory coat or apron, and gloves.

All experiments should first be studied in detail and outlined. A senior chemist or laboratory supervisor should confirm that all details have been understood before the laboratory preparation is started. The nature of each chemical used should be thoroughly understood. The appropriate Material Safety Data Sheet (MSDS) should be examined in detail and then signed off before a chemical is used.

Several **CAUTIONS** are a must to review and are mentioned here again for emphasis:

1. Read all the Material Safety Data Sheets for the raw materials that you plan to use. Contact the supplier if you have any questions.
2. Read all the toxicity information.
3. Use good personal protective equipment such as eye protection, laboratory coats, and gloves (respirators when necessary). The types of gloves and eye protection must be suitable for the operation.
4. Use a well-ventilated hood.
5. Use traps to control toxic vapors or other volatile by-products.
6. Always work with someone else nearby in the same laboratory. *Never work alone!*
7. Read and reread all preparations. Write out equations and understand the chemistry involved.

8. Never scale up experiments unless you are sure this will not lead to a highly exothermic uncontrollable reaction.
9. When uncertain, ask questions of your instructor or supervisor or other qualified professional.

B. WASTE DISPOSAL

For the disposal of waste chemicals, consult your supervisor or the safety director of your institution.

Broken glassware and paper towels used to mop up chemicals should be separated by placing them in special containers or plastic bags. Check with your laboratory instructor or supervisor for the proper guidelines to follow.

1
POLYMERIZATION OF STYRENE AND OTHER OLEFINS

Olefinic and diolefinic monomers can be polymerized using either free-radical, anionic, cationic, or coordination type initiators. These will be discussed individually in each of the sections of this chapter. It is interesting to note that not all monomers respond equally well to each of the types of initiators.

The substituents placed on ethylene greatly affect stereochemistry, resonance, and polarity of the monomer, and have a decided effect on which initiator system works best with it. For example, propene and 1-butene can only be homopolymerized well with coordination catalysts (see Section 4), whereas isobutene responds mainly to cationic initiators. Styrene can be polymerized using any of the four types of initiators. Isoprene and 1,3-butadiene can be homopolymerized with all the initiators except the cationic type, whereas ethylene polymerization can be initiated by all except the anionic type. In the case of vinyl acetate, which has a deficient electron density around the double bond, it can be polymerized mainly by use of a free-radical process. Substituents around the double bond such as chlorine (as in vinyl chloride) will interfere when using cationic or anionic initiators, and it is preferable to polymerize them by a free-radical process.

The free-radical initiating system has the practical advantage that the polymerizations can be carried out in the gas, solid, and liquid phases (bulk, solution, emulsion, suspension, and precipitation techniques). Free-radical reactions can be carried out in water, whereas the other initiators usually require anhydrous conditions.

Many of the anionic acid cationic polymerizations can be considered to have no inherent termination step and may be called "living polymers."

The stereochemistry of the repeating units depends on the structure of the starting monomer, the initiating system, and the conditions of the polymerization reaction. Optical isomerism, geometric isomerism, repeat unit configuration (isotactic, syndiotactic, atactic) and repeat unit orientation (head-to-tail or head-to-head) are some important aspects of the stereochemistry problem. The mechanisms of polymerization (see Eqs. 1–7) will not be discussed here but several worthwhile references should be consulted [1]. This chapter will give mainly examples of some selected preparative methods for carrying out the major methods of polymerization as encountered in the laboratory. All intrinsic viscosities listed in this chapter have units of dl/gm.

This chapter will also present some recent developments in ring-opening methathesis in polymerization.

The free-radical polymerization process which can be carried out in the laboratory is best illustrated by the polymerization of styrene.

Free-radical polymerization processes [2] are carried out either in bulk, solution, suspension, emulsion, or by precipitation techniques. In all cases the monomer used should be free of solvent and inhibitor or else a long induction period will result. In some cases this may be overcome by adding an excess of initiator.

Initiator:

$$I_2 \longrightarrow 2I\cdot \quad (1)$$

Initiation:

$$CH_2=CHR + I\cdot \longrightarrow I-CH_2-\underset{R}{CH}\cdot \quad (2)$$

Propagation:

$$I-\underset{R}{CH_2CH}\cdot + CH_2=CHR \longrightarrow I-CH_2-\underset{R}{CH}-CH_2-\underset{R}{CH}\cdot \xrightarrow{nCH_2=CHR}$$

$$I(CH_2CH)_{n+1}CH_2\underset{R}{CH}\cdot \quad (3)$$

Termination (by radical coupling, disproportionation, or chain transfer):
Radical coupling:

$$\sim\!\!CH_2CH\cdot\ +\ \sim\!\!CH_2\!-\!CH\cdot\ \longrightarrow\ \sim\!\!CH_2CH\ CHCH_2\!\!\sim \qquad (4)$$
$$\qquad\ \ |\qquad\qquad\ \ \ |\qquad\qquad\qquad\ \ \ |\ \ \ \ |$$
$$\qquad\ \ R\qquad\qquad\ R\qquad\qquad\qquad\ R\ \ R$$

Disproportionation of two radicals:

$$\sim\!\!CH_2\!-\!CH\cdot\ +\ \sim\!\!CH_2\!-\!CH\cdot\ \longrightarrow\ \sim\!\!CH_2CH_2\ +\ \sim\!\!CH\!=\!CH \qquad (5)$$
$$\qquad\ \ \ |\qquad\qquad\ \ \ \ |\qquad\qquad\qquad\ \ |\qquad\qquad\ |$$
$$\qquad\ \ \ R\qquad\qquad\ \ R\qquad\qquad\qquad\ R\qquad\qquad R$$

Chain transfer:

$$\sim\!\!CH_2CH\cdot\ +\ R'SH\ \longrightarrow\ \sim\!\!CH_2CH_2\ +\ R'S\cdot \qquad (6)$$
$$\qquad\ \ |\qquad\qquad\qquad\qquad\qquad\ |$$
$$\qquad\ \ R\qquad\qquad\qquad\qquad\qquad R$$

$$R'S\cdot\ +\ CH_2\!=\!CH\ \longrightarrow\ RS'CH_2CH\cdot\ \text{(start of new monomer chain)} \qquad (7)$$
$$\qquad\qquad\qquad\ \ \ |\qquad\qquad\qquad\ \ |$$
$$\qquad\qquad\qquad\ \ \ R\qquad\qquad\qquad R$$

1. BULK POLYMERIZATION

Bulk polymerization consists of heating the monomer without solvent with initiator in a vessel. The monomer–initiator mixture polymerizes to a solid shape fixed by the shape of the polymerization vessel. The main practical disadvantages of this method are the difficulty in removal of polymer from a reactor or flask and the dissipation of the heat evolved by the polymerization. A typical example is shown in Preparation 1-1. This method finds importance in producing cast or molded products, such as plastic scintillators, in small or very large shapes, but it is difficult since the formation of local hot spots must be avoided.

In the use of polystyrene, the polymerization reaction is exothermic to the extent of 17 Kcal/mole or 200 BTU/lb (heat of polymerization). The polystyrene produced has a broad molecular weight distribution and poor mechanical properties. The residual monomer in the ground polymers can be removed by use of efficient devolatilization equipment. Several reviews are worth consulting [3].

The bulk polymerization of styrene to give a narrow molecular weight distribution has appeared in a U.S. patent [4]. The polydispersity reported was 2.6 at a 93% conversion and had a number average molecular weight of about 100,000. This was accomplished by polymerizing styrene in the presence of 1.0% of 4-*tert*-butylpyrocatechol at 127°C for 2.27 hr. Heating in the absence of the latter gave a polydispersity of 3.3 with a number average molecular weight of 79,200.

Several references to the bulk polymerization of styrene are worth consulting [5]. Most consider a continuous bulk polymerization apparatus with some using spraying of the monomer through a nozzle. The controlled evaporation of unreacted monomer is one method of removing the heat of reaction.

1-1. Preparation of Polystyrene by the Bulk Thermal Polymerization of Styrene [6,7]

$$n \, \underset{}{\text{CH=CH}_2\text{-C}_6\text{H}_5} \longrightarrow [\text{-CH-CH}_2\text{-}]_n \text{(C}_6\text{H}_5) \tag{8}$$

To a polymer tube is added 25 gm of distilled styrene monomer (preferably inhibitor-free) (see Notes a, b), then the tube is flushed with nitrogen, sealed, and placed in an oil bath at 125°C for 24 hr. At the end of this time the monomer is converted in 90% yield to the polymer. Heating for 7 days at 125°C and then 2 days at 150°C gives a 99% or better conversion to polymers (see Note c). If all the air has been excluded the polystyrene will be free of yellow stains on the surface [7]. The polystyrene is recovered by breaking open the tube and can then be purified by dissolving in benzene and precipitating it with methanol. The solid polymer is filtered, and dried at 50°–60°C in a vacuum oven to give a 90% yield (22.5 gm). The molecular weight is about 150,000–300,000 as determined by viscometry in benzene at 25°C. (**CAUTION:** Benzene is carcinogenic. Read all MSDS.)

NOTES: (a) The monomer should be distilled prior to use to give a faster rate of polymerization. If this is not practical then the sample should be washed with 10% aq. NaOH and dried. (b) Styrene monomer also polymerizes at room temperature over a period of weeks or months. (c) Use of 0.5% of benzoyl peroxide allows one to prepare the polymer at 50°C over a 72 hr period to give $\eta_i = 0.4$ (0.5 gm/100 ml C_6H_6 at 25°C). This type of combination is suitable for preparing castings or molds of polystyrene objects.

Additional information on the polymerization of styrene and related monomers has been described by Boundy and Boyer [8] and also by the Dow Chemical Company [9].

The thermal polymerization of styrene at 150°C also gives a mixture of dimers of which 11 C_{16} hydrocarbons have been identified by Mayo [10].

2. SUSPENSION POLYMERIZATION

In suspension polymerization a catalyst is dissolved in the monomer, which is then dispersed in water. A dispersing agent is added separately to stabilize the resulting suspension. The particles of the polymer are 0.1 to 1 mm in size. The rate of polymerization and other characteristics are similar to those found in bulk polymerization. Some common dispersing agents are polyvinyl alcohol, polyacrylic acid, gelatin, cellulose, and pectins. Inorganic dispersing agents are phosphates, aluminum hydroxide, zinc oxide, magnesium silicates, and kaolin. Several worthwhile references should be consulted for more details [11–13].

Suspension polymerization leading to the control of the size and size distribution of polymer particles is more of an art than a science. Much of the data for commercial processes appear in the patent literature. A recent review by Guyot [14] with emphasis on polymer supports is worth consulting [15].

The design of the reactor and the stirrer plays an important role in governing the shear distribution inside the reactor, which then affects the polymer particle size. The other important factors that affect polymer size and distribution are ratio of monomer to water or solvents, the temperature of the reaction, the type of initiator/ level used, and the suspending agent.

Some reports, worth referring to, describe the effects of the type of water-soluble polymer used as the suspending agent and its results on the particle size of the polymer being prepared from the monomer [16]. Additional references on the parameters important in suspension polymerization refer to influences on particle size effects [17], the effect of the apparatus used [18], initiators [19], stirring efficiency [20], dispersants [21], and electrolytes [22].

The preparation of larger sized polymer particles of polystyrene that are also monodisperse are reported to be prepared by a special technique involving seeded suspension polymerization [22]. In addition the effect of using ultrasonic irradiation for the suspension polymerization of styrene has been reported [24].

1-2. Preparation of Polystyrene by Suspension Polymerization [13]

To a 350 ml glass bottle is added 100 gm of styrene (see Note), 0.3 gm of benzoyl peroxide, 150 ml of water, and 3.0 gm of zinc oxide. Then conc. aqueous ammonium hydroxide is added to adjust the pH to 10. The bottle is then sealed and rotated (30 rpm) in an oil bath for 7 hr at 90°C and 5 hr at 115°C. At the end of this time the pH is 7.25. The polymer granules are filtered, suspended in water, and acidified to pH approx. 2 with 10% aqueous hydrochloric acid in order to remove the zinc oxide. The polymer granules are then washed with water and dried.

NOTE: This process works well for other olefins such as vinyltoluene, vinylxylene, and *tert*-butylstyrene.

3. EMULSION POLYMERIZATION

The system basically consists of water and 1–3% of a surfactant (sodium lauryl sulfate, sodium dodecyl benzenesulfonate, or dodecylamine hydrochloride) and a water-soluble free-radical generator (alkali persulfate, hydro-peroxides, or hydrogen peroxide–ferrous ion). The monomer is added gradually or is all present from the start. The emulsion polymerization is usually more rapid than bulk or solution polymerization for a given monomer at the same temperature. In addition the average molecular weight may also be greater than that obtained in the bulk polymerization process. The particles in the emulsion polymerization are of the order of 10^{-5} to 10^{-7} m in size. It is interesting to note that the locus of polymerization is the micelle and only one free radical can be present at a given time. The monomer is fed into the locus of reaction by diffusion through the water where the reservoir of the monomer is found. If another radical enters the micelle then termination results because of the small volume of the reaction site. In other words, in emulsion polymerization the polymer particles are not formed by polymerization of the original monomer droplets but are formed in the micelles to give polymer latex particles of very small size. For a

1 POLYMERIZATION OF STYRENE AND OTHER OLEFINS 7

review of the roles of the emulsifier in emulsion polymerization see a review by Dunn [25]. Other references with a more detailed account of the field should be consulted [26, 27]. These references are only a sampling of the many that can be found in *Chemical Abstracts*.

The polymer in emulsion polymerization is isolated by either coagulating or spray-drying.

In some cases the latex is the desired product, as in the use of latex paint formulations. For this application a multicomponent emulsifier system and stabilizers are used.

Mercaptan chain transfer agents are used in the emulsion polymerization of styrene to control the molecular weight of the resulting polymers. Varying the concentration of the initiator and monomer had a negligible effect on the chain transfer reaction. Some of the mercaptans investigated are n-decyl mercaptan, n-dodecyl mercaptan, and t-dodecyl mercaptan [28].

1-3. Emulsion Polymerization of Styrene [29]

To a resin kettle equipped with a mechanical stirrer, condenser, and nitrogen inlet tube is added 128.2 gm of distilled water, 71.2 gm of styrene, 31.4 ml of 0.680% of potassium persulfate, and 100 ml of 3.56% soap solution (see Note). The system is purged with nitrogen to remove dissolved air. Then the temperature is raised to 80°C and kept there for about 3 hr to afford a 90% conversion of polymer. The polymer is isolated by freezing–thawing or by adding alum solution and boiling the mixture. The polystyrene is filtered, washed with water, and dried.

NOTE: In place of soaps such as sodium stearate one can use 1% by wt. of either sodium dodecyl benzenesulfonate or sodium lauryl sulfate based on the weight of the monomer.

Robb [30] has studied the emulsion polymerization of styrene at 40°C [styrene (30 gm), water (300 gm), sodium dodecyl sulfate (2.1 gm), and potassium persulfate (0.60 gm)]. After 2 hr a 92% conversion was obtained. Robb determined the number of particles per unit volume of latex during the emulsion polymerization of styrene and described their properties.

4. ANIONIC POLYMERIZATIONS

The anionic polymerization of styrene was first reported in 1914 by Schlenk and co-workers [31] and reinvestigated by Szwarc [32] and others [33]. More recently, Priddy reported that the anionic polymerization of styrene is industrially feasible. A process to produce a broad molecular weight distribution of polystyrene [$\overline{M}_w/\overline{M}_n => 2.0$) in a continuously stirred tank reactor (CSTR) at 90–110°C has been described [34]. The polymers also have excellent color and are purer than those formed in the free-radical polymerization process. The use of α-methylstyrene comonomer gave increased heat resistance [35]. Almost 60 years ago sodium and lithium metal were used to polymerize conjugated dienes such as butadiene [36–38], isoprene [36], 1-phenylbutadiene [37], and 2,3-dimethylbutadiene [39]. Ziegler [40] in 1929 described the addition of organoalkali compounds to a double bond. In 1940 the use of butyllithium for the low-pressure polymerization of ethylene was described [41]. In 1952 the kinetics of the anionic polymerization of styrene using KNH_2 was reported [42]. Some anionic polymerizations have been described as living polymers in the absence of impurities [43].

Electron-withdrawing substituents adjacent to an olefinic bond tend to stabilize carbanion formation and thus activate the compound toward anionic polymerization [44].

The relative initiator activities are not always simple functions of the reactivity of the free anion but probably involve contributions by complexing ability, ionization, or dissociation reactions [45].

Waack and Doran [46] reported on the relative reactivities of 13 structurally different organolithium compounds in polymerization with styrene in tetrahydrofuran at 20°C. The reactivities were determined by the molecular weights of the formed polystyrene. The molecular weights are inversely related to the activity of the respective organolithium polymerization initiators. Reactivities decreased in the order alkyl > benzyl > allyl > phenyl > vinyl > triphenylmethyl.

The structure–reactivity behavior found for similar organosodium polymerization initiators of styrene [47] or that for addition reactions with 1,1-diphenylethylene [48] is almost identical with that found for the lithium initiators of Table I. It is interesting to note from Table I that the reactivity of lithium naphthalene, a radical

1 POLYMERIZATION OF STYRENE AND OTHER OLEFINS

TABLE I
"STANDARD POLYMERIZATIONS" OF STYRENE
IN TETRAHYDROFURAN SOLUTION AT 20°C

Organolithium catalyst	Mol. wt. of polymer^a (temp, °C)
t-Butyl	3,200^b (−66)
	3,200 (−40)
sec-Butyl	3,500^b (−69)
Ethyl	3,500
n-Butyl	3,600
α-Methylbenzyl	3,700
Crotyl	6,500
Benzyl	6,700
Allyl	9,600
p-Tolyl	9,900
Phenyl^c	12,000
Phenyl^d (LiX)	24,000 (LiCl)
	22,000 (LiBr)
Methyl	19,000
Vinyl	23,000
Triphenylmethyl^e	66,000
Triphenylmethylsodium	53,000
Lithium naphthalene	6,000

^{a}Average values.
^{b}At higher temperatures there is rapid reaction with THF.
^{c}Salt-free.
^{d}Contains equimolar lithium halide.
^{e}Contains equimolar LiCl. Lithium halides are indicated to have little effect on the reactivity of such resonance stabilized species. Reprinted from R. Waack and M. A. Doran, *J. Org. Chem.* **32**, 3395 (1967). Copyright 1967 by the American Chemical Society. Reprinted by permission of the copyright owner.

anion type initiator, is between that of the alkyl lithiums and the aromatic lithium initiators.

The anionic polymerization of styrene using the organolithium initiators can be described as a termination-free polymerization, as shown in Eqs. (9) and (10).

$$\text{RLi} + \text{St} \longrightarrow \text{RStLi} \quad (9)$$

$$\text{RStLi} + n\text{St} \longrightarrow \text{RSt}_n\text{StLi} \quad (10)$$

The degree of polymerization (DP) is determined by the ratio of the overall rate of propagation to that of initiation.

"Living" polymers have been extensively investigated by Szwarc and coworkers [49] who have shown that the formed polymer can spontaneously resume its growth on addition of the same or different fresh monomer. Block copolymers are easily synthesized by this technique. The lack of self-termination is overcome by the addition of proton donators, carbon dioxide, or ethylene oxide.

The living nature of the poly(styryl) anion allows one to prepare block copolymers with a great deal of control of the block copolymer structure. The preparation of diblock, triblock, and other types of multiblock copolymers has been reviewed [50]. Several of these block copolymers are in commercial use. The basic concept involves first preparing polystyrene block ($RSt_n StLi^+$—see Eq. 10) and then adding a new monomer which can be added to start another growing segment.

The living anionic ends can be functionalized by adding such agents as ethylene oxide, carbon dioxide, methacryloyl chloride, etc. [51] (see Eq. 11). The resulting new polymer is capable of being copolymerized with additional monomers. This process can lead to the formation of various graft copolymers [50].

$$RSt_n St^- \xrightarrow{O} RSt_n StCH_2CH_2O^- \tag{11}$$

with reagents: $H_2C=C(CH_3)-COCl$, CH_3OH, CO_2

giving products: $Rst_n StC(=O)-C(CH_3)=CH_2$, $RSt_n StCOO^-$, $RSt_n StH$

The use of sodium and naphthalene complex in an ethereal solvent to initiate styrene polymerization involves an electron transfer mechanism. Initiation occurs by a three-step process: (1) electron transfer from sodium to naphthalene, (2) electron transfer from naphthalene radical anion to the styrene double bond, and (3) recombination of the styrene radical anions leading to the dianion of styrene dimer which is the initiating species [52, 53] (see Eq. 12).

1 POLYMERIZATION OF STYRENE AND OTHER OLEFINS

$$\text{naphthalene} + Na \rightleftharpoons [\text{naphthalene}]^{\cdot -} + Na^+ \xrightarrow{CH_2=CH-Ph}$$

$$\longrightarrow \left(\begin{array}{c} CH^{\cdot}-CH_2^- \\ Ph \end{array} \right) Na^+ + \text{naphthalene} \longrightarrow Na^+ \left(\begin{array}{c} CH^- -CH_2-CH_2-CH^- \\ Ph \hspace{2cm} Ph \end{array} \right) Na^+ \quad (12)$$

The propagation of this polymerization then proceeds from both ends. The solvent is very important in the rates of anionic polymerization as well as the type of counter ion used [54].
Reference [55] is a general review of the anionic polymerization process and is worth consulting.

1-4. Preparation of Polystyrene by the Polymerization of Styrene Using Sodium Naphthalene Catalyst [53]

$$\text{Styrene} + \text{naphthalene}^{\cdot -} \longrightarrow \text{styrene}^{\cdot -} \xrightarrow{\text{styrene}} \text{polystyrene}^{\cdot -} \quad (13)$$

$$\text{Polystyrene}^{\cdot -} \begin{array}{c} \xrightarrow{CH_3OH} \text{polystyrene} \\ \xrightarrow{\text{styrene}} \text{further polymerization} \end{array} \quad (14)$$

(a) Preparation of Sodium Naphthalene. A 2-liter three-necked flask, which has been dried at 150°C and then assembled warm with a nitrogen gas stream going through it, is equipped with a Teflon-sealed stirrer (or ground glass type).

$$\text{naphthalene} + Na \xrightarrow{THF} Na^+ \left(\text{naphthalene}^{\cdot -}\right) \quad (15)$$

Then 1 liter of a molal solution of naphthalene in pure dry tetrahydrofuran is added followed by 25 gm of sodium sticks (2–3 cm long × 3–5 cm square on the end). After adding the sodium the mixture is stirred rapidly at the start. The reaction starts in 1–3 min and some cooling may be necessary to keep the temperature from exceeding 30°C. At 20°–25°C the reaction takes about 2 hr, and then an aliquot of the mixture is added to methanol and titrated with dilute standard HCl using methyl red.

(b) Polymerization of styrene [53]. To a flask as described previously but with a side arm with a serum cap is added 60 ml of pure dry THF containing 3.3×10^{-4} mole of sodium naphthalene (positive nitrogen pressure maintained throughout reaction). Then the contents are cooled to $-80°C$ and 9.2 gm of dry pure distilled styrene is injected. Polymerization proceeds rapidly and the reaction mixture turns a bright red color due to the formation of the styrene anion. After completion of the reaction the solution is warmed to room temperature and the viscosity (see Note) is 1.2–1.5 sec. The solution is recooled to $-80°C$, and an additional 7.7 gm of styrene is 50 ml of THF is injected. After 1–2 hr the color of the reaction is still bright red and the viscosity (see Note) of the reaction mixture is 18–20 sec at room temperature. The polymerization is quenched (red color disappears) by the addition of cold methanol, filtered, washed with methanol, and dired to afford 16.6 gm (98%) of polystyrene, inherent viscosity in toluene at 0.5% concentration is 1 to 1.5. This preparation proves that the living ends of the polystyrene were able to initiate further polymerization when the second addition of styrene was made.

NOTE: Viscosity taken without removing sample from reaction vessel. Szwarc [53] has shown that styrene initiated as described above can also react with isoprene to give block copolymers.

5. PREPARATION OF MONODISPERSED MICROSPHERES OF POLYSTYRENE

The polymerization of styrene to produce uniform microspheres with diameters of 1 μm or larger in a single step has been a challenge. The use of conventional suspension polymerization techniques is unsatisfactory since they invariably lead to a very broad size distribution. The use of dispersion polymerization methods, i.e., polymerizations in media with little or no water, is promising, albeit many interrelated factors have to be controlled for success. Reference [56] reviewed this field and cites at least 16 leading references.

Rudin and co-workers [56] make use of a clear solution of ethanol, styrene, poly(vinyl pyrrolidone) [PVP], and AIBN. Upon initiation at 70°C, a homopolymer of styrene and a graft copolymer of polystyrene and PVP form alongside of each other. As the polystyrene chains grow initially, they gradually reach their solubility limit in the ethanol solution. Several chains then coalesce to form unstable nuclei. Along with this step, some of the graft copolymer chain are also adsorbed until the graft copolymer stops further adsorption of polystyrene chains and a fixed number of stable nuclei form. This nucleation step is short in duration and is followed by a relatively lengthy growth stage to form monodispersed particles. Since the objective of this work was the production of microspheres for peptide syntheses, some commercial divinylbenzene was incorporated in the preparative procedure. The resulting particles were said to be resistant to the solvent systems used in peptide synthesis. The uniformity of particle diameters produced by this method, as evidenced by the large standard deviations, is not up to the quality usually associated with monodispersed microspheres.

1-5. Monodispersed Crosslinked Microspheres by Dispersion Polymerization [56]

To a 1 liter glass resin flask equipped with a stainless steel stirrer which was sealed with a Teflon gasket, nitrogen inlet, addition funnel, condenser, and thermometer and maintained at 70°C with a water bath, was charged, with stirring at approximately 100 rpm, 270.0 gm of ethanol, 7.2 gm of PVP (mol. wt. 40,000), 0.364 gm of uninhibited divinylbenzene (derived from a commerical 55% solution of the mixed *meta* and *para* isomers), and 75 gm of uninhibited styrene. The reaction mixture was stirred under nitrogen for 30 min. Then a solution of 2 gm of AIBN in 25 gm of styrene and 20 gm of ethanol was added. Heating and stirring was continued for 18 hr. After cooling, the product was filtered through cheese cloth to remove some coagula. From the Coulter Multisizer, the particle size (reported as a d_{50}) was 4.57 μm with a "geometric standard deviation" [calculated as $(d_{84}/d_{16})^{1/2}$] of 1.02.

Reference [56] gives many details of the effect of reaction variables such as the temperature at which the AIBN is added; the effects of different levels of AIBN, PVP, and styrene on particle size; the effect of adding either toluene or water to the reaction medium; the addition of water initially or after nucleation; and the effect of the level of divinylbenzene and the method of its addition.

6. POLYMERIZATION OF OTHER OLEFINS USING A CATIONIC POLYMERIZATION PROCESS

Cationic polymerization has a history dating back about 183 years and has been extensively investigated by Plesch [57], Dainton [58], Polanyi [59], Pepper [60, 61], Evans [62], Heiligmann [63], and others [64]. Whitmore [65] is credited with first recognizing that carbonium ions are intermediates in the acid-catalyzed polymerizations of olefins. The recognition of the importance of proton-donor cocatalysts for Friedel–Crafts catalysts was first reported by Polanyi and co-workers [59]:

$$MX_n + SH \rightleftharpoons [MX_n S]^- H^+ \quad (16)$$

(SH and RX = Lewis base)

$$MX_n + RX' \rightleftharpoons [MX_n R]^- E^+ \quad (17)$$

Some common initiators for cationic polymerization reactions are either protonic acids, Friedel–Crafts catalysts (Lewis acids), compounds capable of generating cations, or ionizing radiation.

Of all the acid catalysts used [66–68] sulfuric acid is the most common. Furthermore, sulfuric acid appears to be a stronger acid than hydrochloric acid in nonaqueous solvents. Some other commonly used catalysts are HCl, H_2SO_4, BF_3, $AlCl_3$, $SnCl_4$, $SnBr_4$, $SbCl_3$, $BeCl_3$, $TiCl_4$, $FeCl_3$, $ZnCl_2$, $ZrCl_4$, and I_2. The use of the Lewis acid catalysts requires traces of either a proton donor (water) or a cation donor (a tertiary amine hydrohalide) effectively to initiate the polymerization process. For example, in the absence of the latter, rigorously dry systems cannot be used to initiate the cationic polymerization of isobutylene [69].

For alkenes the reactivity is based on the stability of the carbonium ion formed and they follow the order: tertiary >

1 POLYMERIZATION OF STYRENE AND OTHER OLEFINS

secondary > primary. Thus olefins react as follows:
$(CH_3)_2C=CH_2 \simeq (CH_3)_2C=CHCH_3 > CH_3CH=CH_2 > CH_2=CH_2$. Allylic and benzylic carbonium [70, 71] ions are also favored where appropriate.

The cationic polymerization process has been reviewed and several references are worth consulting [60, 72, 73].

Certain aluminum alkyls and aluminum dialkyl halides in the presence of proton- or carbonium-donating cocatalysts act as effective polymerization catalysts.

In the absence of monomers, trimethylaluminum (0.5 mole) reacts with t-butyl chloride (1.0 mole) at $-78°C$ to give a quantitative yield of neopentane [73]. Kennedy found that aluminum trialkyls ($AlMe_3$, $AlEt_3$, $AlBu_3$) in the presence of certain alkyl halides are efficient initiators for the cationic polymerization of isobutylene, styrene, etc. [74].

$$AlR_3 + R'X \longrightarrow [AlR_3X]^- + [R]^+ \qquad (18)$$

For example, a flask containing 10 gm of isobutylene and 20 gm of carbon tetrachloride is cooled to $-78°C$ and boron trifluoride gas is bubbled into the mixture. The isobutylene polymerizes and the whole mass apparently solidifies. On warming to room temperature the carbon tetrachloride melts away, leaving 5.0 gm (50%) of polymer, mol. wt. 20,000.

NOTE: Ethylene dichloride can also be used in place of carbon tetrachloride, and other suitable solvents for the polymerization of isobutylene are either ethylene dichloride or a mixture of 2 volumes of liquid ethylene and 1 volume of ethyl chloride. Propane is not as good as the latter solvents but gives a higher molecular weight product than for the case of carbon tetrachloride above.

A good account of the early development of high polymers and copolymers of isobutylene is given by Schildknecht [75].

Thomas, Sparks, and Frolich have also described the early work and patents on the polymerization of isobutylene with acidic catalysts ($TiCl_4$, $AlCl_3$, BF_3) [76]. In particular, Thomas and Sparks described the early work on the $AlCl_3-CH_3Cl$-catalyzed copolymerization of isobutylene with isoprene at low temperatures and gave several worthwhile examples in their patent [77]. Additional examples were later described by Calfee and Thomas [78].

1-6. Preparation of Butyl Rubber (Copolymerization of Isobutylene with Isoprene Using BF_3 Catalyst) [79]

$$CH_2{=}C(CH_3)_2 + CH_2{=}CH-\underset{\underset{CH_3}{|}}{C}{=}CH_2 \xrightarrow[-78°C]{BF_3}{CH_3Cl}$$

$$\left[CH_2-C(CH_3)_2-CH_2-CH{=}\underset{\underset{CH_3}{|}}{C}-CH_2 \right]_n \quad (19)$$

To an oven-dried micro resin flask equipped with a thermocouple, stirrer, and cooled jacketed dropping funnel and placed in a dry ice–acetone bath ($-78°C$) is added 9.7 ml (6.8 gm, 0.121 mole) of isobutylene and 0.3 ml of isoprene dissolved in 30 ml of methyl chloride (all volumes measured at $-78°C$). Then BF_3 (1.8×10^{-5} mole) dissolved in methyl chloride is added dropwise so that a temperature rise no higher than $1°$ is experienced. After 1–2 hr reaction at $-78°C$ the polymerization is terminated by adding 10 ml of precooled methanol and stirring for 5 min. The unreacted monomers are evaporated; the polymer is washed with methanol and dried for 48 hr at 45–50°C to afford 4.8 gm (68.6%), mol. wt. 140,200; DP = 2510. The mole percent isoprene content is 1.84 and is determined as described in Refs. [80, 81].

Kennedy has also described the preparation of butyl rubber utilizing an AlR_2X catalyst with an HX promoter at $0°$ to $-100°C$ [81]. In this case R = C_1 to C_{12} aliphatic hydrocarbon radical and X = F, Cl, or Br.

Marvel and co-workers [82] have also described the cationic polymerization of butadiene using one of several Friedel–Crafts catalysts at various temperatures and concentrations. Aluminum chloride and chlorosulfonic acid were effective at $-75°C$ and stannic chloride, boron trifluoride etherate, boron trifluoride hydrate, sulfuric acid, and fuming sulfuric acid required higher temperatures to bring about substantial polymerization. The catalysts usually were added in chloroform or ethyl bromide as solvent. Butadiene also was found to copolymerize well with styrene at $-75°C$ using aluminum chloride in ethyl bromide as catalyst.

7. POLYMERIZATION OF OTHER OLEFINS BY AN ANIONIC POLYMERIZATION PROCESS

1-7. Preparation of Polybutadiene by the Amylsodium-Catalyzed Polymerization of Butadiene [83]

$$n\text{CH}_2=\text{CH}-\text{CH}=\text{CH}_2 \xrightarrow[\text{pentane}]{\text{C}_5\text{H}_{11}\text{Na}} +\text{CH}_2-\text{CH}=\text{CH}_2-\text{CH}_2+_n \quad (20)$$

A 500 ml creased Morton flask that has been dried at 150°C is assembled warm with a mechanical stirrer, dropping funnel, and nitrogen inlet-outlet. While a positive nitrogen atmosphere is maintained a 300 ml pentane suspension of amylsodium (see Note) (0.083 mole) is added. The flask is cooled to 0°–10°C and 1.00 mole of butadiene is added over a 1 hr period. After stirring for an additional hour the reaction mixture is carbonated by pouring the contents of the flask on solid carbon dioxide. When the solid carbon dioxide evaporates 200 ml of water is dropwise added. Then 0.4 gm of hydroquinone is added and the pentane is removed by warming on a steam bath. The polymer is removed from the aqueous layer, acidified with hydrochloric acid, and steam-distilled in order to remove any decane. The rubbery mass is dried by distilling chloroform from it and then dried under reduced pressure to afford 45.4 gm (74%).

NOTE: Amylsodium is prepared in 75–80% yield by the dropwise addition (1 hr) of the halide (0.25 mole) to two equivalents of finely divided sodium in pentane at −10°C.

n-Butyllithium in *n*-pentane or hexane can be used in place of amylsodium as the catalyst.

8. CATALYTIC BREAKTHROUGHS IN POLYMERIZING OLEFINS

A. COORDINATION CATALYST POLYMERIZATIONS

The earliest reported coordination polymerization was the cationic polymerization of vinyl isobutyl ether [84]. Other early references reported the polymerization of ethylene using titanium tetrachloride–aluminum–aluminum chloride [85] and the polymerization of styrene using phenylmagnesium bromide–Ti(OBu)$_4$ [86].

Ziegler [87] first recognized the importance of the catalyst and the novel type of polymerization it induced, especially for ethylene monomer. The polyethylene produced was very linear (m.p. 135°C), whereas the high-pressure radical process gives a short-chain branched product (m.p. 120°C). Natta [88] almost at the same time reported on a similar catalyst useful for a wide range of olefin, diolefin, and acetylenic monomers. These new catalysts have now become known as the Ziegler–Natta catalysts. These catalysts are complexes formed by the reaction of halides or other derivatives of transition metals of groups IV–VIII and alkyls of metals of groups I–III, for example, as shown in Eq. (21).

$$AlR_3 + MCl_n \longrightarrow R_2AlCl + RMCl_{n-1} \qquad (21)$$

The catalysts, which are based on insoluble transition metal compounds or complexes, afford little information as to the active species present. Many of the catalysts have metal salts in varying oxidation states. These catalysts therefore have a structure which is not known with certainty and leads to difficulties in trying to reproduce results.

Recently various well-defined π-allylnickel halides [89, 90] have been reported to initiate diene polymerization. However, the polymerization is slow and also affords low molecular weight products.

Cooper has reviewed the coordination catalyst mechanism of polymerization of conjugated dienes and his reference is worth consulting [91]. Other earlier reviews, papers, and texts should also be consulted [92].

Recently Grubbs [93] has described new organometallic catalysts for the polymerization of cyclic-olefins. This process is described as a ring-opening metathesis polymerization (ROMP) with a "carbene"-like reagent such as (I), shown below [94]:

$$Cp_2TiCl_2 + 2AlMe_3 \longrightarrow CH_4 + Cp_2Ti-CH_2AlClMe_2-AlMe_2Cl$$

(I)

$$(I) = \underset{\text{Cp}_2\text{Ti}}{\overset{\text{Cl}}{\diagdown}} \underset{\text{CH}_2}{\overset{}{\diagup}} Al(CH_3)_2 = [Cp_2Ti=CH_2] \cdot AlMe_2Cl \qquad (22)$$

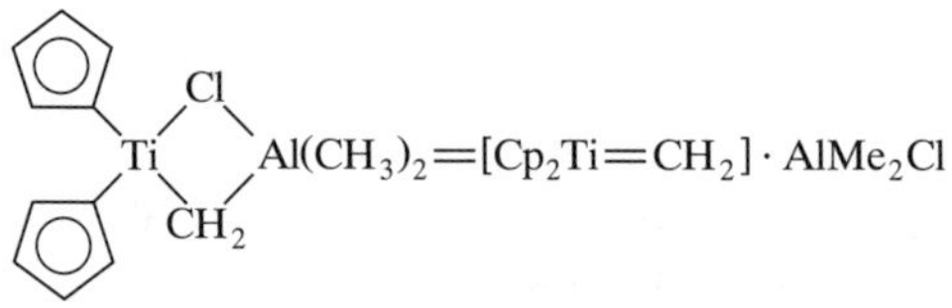

(Tebbe reagent)[94]

where Cp = cyclopentadionyl

1 POLYMERIZATION OF STYRENE AND OTHER OLEFINS

Gilliom and Grubbs [95] used (I) to react with norbornene to afford a ring-opened polynorbornene with a *cis* to *trans* ratio of 38:62, which can also be classified as a "living" polymerization.

$$(I) + \text{[norbornene]} \xrightarrow{N,N\text{-Dimethylaminopyridine}} \tag{23}$$

$$Cp_2Ti\text{[structure]} \xrightarrow{65°C} \text{[polymer]}_n$$

B. METALLOCENE CATALYST POLYMERIZATIONS

Dow recently announced [96, 97] its INSITE process, which utilizes metallocenes to polymerize ethylene and to copolymerize ethylene with styrene. The latter was said to have properties similar to flexible PVC. Dow also announced that it has converted about 1 billion pounds of production of polyethylene to their INSITE metallocene catalyst process.

Other chemical companies such as Exxon and Union Carbide have reported [98] related metallocene technology ventures in the polyethylene area.

The active metallocene catalyst used by Dow may have the following generalized structure:

$$\begin{array}{c} CH_3 \\ \diagdown \\ CH_3 \end{array} Si \begin{array}{c} \diagup \\ \diagdown \end{array} \begin{array}{c} N-Te^+-R \\ | \\ X^- \end{array} \quad \begin{array}{c} CH_3 \\ CH_3 \diagdown \diagup CH_3 \\ \text{[Cp ring]} \\ CH_3 \diagup \diagdown CH_3 \end{array} \quad \text{where } X^- = B\left(-\underset{CF_3}{\overset{CF_3}{\bigcirc}}-\right)_4^- \tag{24}$$

For other references to this area see Refs. [99, 100].

REFERENCES

1. C. Walling, "Free Radicals in Solution," Wiley, New York, 1957; J. C. Bevington, "Radical Polymerization," Academic Press, New York, 1961; P. E. M. Allen and P. H. Plesch, in "The Chemistry of Cationic Polymerization" (P. H. Plesch, ed.), Macmillian, New York, 1963; P. J. Flory, "Principles of Polymer Chemistry," Cornell Univ. Press, Ithaca, New York, 1953; R. W. Lenz, "Organic Chemistry of Synthetic High Polymers," Wiley (Interscience),

New York, 1967; C. L. Arcus, *Progr. Stereochem.* **3**, 264 (1962); M. L. Huggins, G. Natta, V. Derreux, and H. Mark, *J. Polym. Sci.* **56**, 153 (1962); G. Natta, L. Porri, P. Corradini, G. Zanini, and F. Ciampelli, *ibid.* **51**, 463 (1961); N. Beredjick and C. Schuerch, *J. Amer. Chem. Soc.* **78**, 2646 (1956); C. S. Marvel and R. G. Wollford, *J. Org. Chem.* **25**, 1641 (1960); G. B. Butler and R. W. Stackman, *ibid.*, p. 1643 (1961); M. Farina, M. Peraldo, and G. Natta, *Angew. Chem., Int. Ed. Engl.* **4**, 107 (1965); J. E. McGrath, *J. Chem. Educ.* **58**, 844 (1981); American Chemical Society Course on "Polymer Chemistry: Principles and Practice" at Virginia Tech., Blacksburg, Virginia, March, August, and December, 1989 (Instructors: Professors M. E. McGrath, T. C. Ward, and G. L. Wilkes).
2. J. A. Faucher and E. P. Reding, *in* "Crystalline Olefin Polymers" (A. V. Raff and K. W. Doak, eds.), Chapter 13, Wiley (Interscience), New York, 1965.
3. J. L. Amos, *Polym. Eng. Sci.* **14**, 1 (1974); R. F. Boyer, *J. Macromol Sci. Chem.* **15**, 1411 (1981); R. H. M. Simon and D. C. Chappelear, *in* J. N. Henderson and T. C. Bonton, eds., "Polymerization Reactions and Processes," ACS Symposium Series, **104**, 71–112, American Chemical Society, Washington, D.C. (1979).
4. R. A. Hall and J. I. Rosenfeld, U.S. Patent 4,713,421 (12/15/87); *Chem. Abstr.* **108**, 205285s (1988).
5. S. Omi, I. Iwata, K. Innbuse, M. Isu, and M. Suku, *Int. Polymer Process* **2** (3–4), 198 (1988); *Chem. Abstr.*, **108**, 151108r (1988); T. Uetake, T. Kaino, and T. Yushizawa, *Jpn. Kokai Tokkyo Koho* JP 60/152506A2 (8/10/85); *Chem. Abstr.* **104**, 110759c (1986); K. T. Nguyen, E. Flaschel, and A. Renken, *Chem. Eng. Commun.* 36 (1–6), 251 (1985); *Chem. Abstr.* **103**, 24013w (1985); B. M. Baysal, E. Bryramli, H. Yuruk, and B. Hazer, *Macromol. Chem.* **6** (6) 1269 (1985); *Chem. Abstr.* **103**, 71761c (1985); J. L. McCurdy, *Review Modern Plastics* **39** (285), 309–310, 315–316 (1980); *Chem. Abstr.* **92**, 21609o (1980).
6. Author's Laboratory (S.R.S.); S. R. Sandler, P. J. McGonigal, and K. C. Tsou, *J. Phys. Chem.* **66**, 166 (1962).
7. S. R. Sandler, *Int. J. Appl. Radiat. Isotop.* **16**, 473 (1965).
8. R. H. Boundy and R. F. Boyer, eds., "Styrene—Its Polymers Copolymers and Derivatives," Amer. Chem. Soc., Monogr. No. 115. Van Nostrand-Reinhold, Princeton, New Jersey, 1952.
9. Dow Chemical Company, "Styrene-Type Monomer," Technical Brochure No. 114-151-69 (1969); Dow Chemical Co., "Storage and Handling of Styrene-Type Monomers," Technical Brochure No. 170-280-6M-1067 (1967).
10. F. R. Mayo, *J. Amer. Chem. Soc.* **90**, 1289 (1968).
11. G. C. Eastmond, *Encycl. Polym. Sci. Technol.* **7**, 361 (1967).
12. E. Farber, *Encycl. Polym. Sci. Technol.* **13**, 522 (1970); G. J. Gammon, C. T. Richards, and M. J. Symes, British Patent 1,243,057 (1971); H. Nishikawa *et al.*, Japanese Patent 71/06,423 (1971); K. Wilkinson, British Patent 1,226,959 (1971); W. N. Maclay, *J. Appl. Polym. Sci.* **15**, 867 (1971); J. R. Hiltner and W. F. Bartoe, U.S. Patent 2,264,376 (1941); A. H. Turner and W. D. Bannister, British Patent 873,948 (1960); Kanegafuchi Chemical Industry Co., Ltd., French Patent 1,373,240 (1964); H. K. Chi, U.S. patent 3,258,453 (1966); Shell

Internationale Research Maatschappij N. V., French Patent Appl. 2,020,845 (1970).
13. Dow Chemical Company, Netherlands Patent Appl. 6,514,851 (1966).
14. A. Guyot in "Syntheses and Separation Using Functional Polymers" (D. C. Sherrington and P. Hudge, eds.), pp. 1–43. John Wiley & Sons, Ltd., New York, 1988.
15. J. A. Patterson in "Biological Aspects of Reactions on Solid Support" (G. R. Stark, ed.), pp. 189–213, Academic Press, New York, 1971.
16. H. Hopff, H. Lussi, and E. Hammer, *Makromol. Chem.*, **82**, 185 (1965); D. C. Sherrington, *Brit. Polym. J.*, **16**, 164 (1984); C. Jouitteau, A. Revillon, and A. Guyot, unpublished results; M. Tomoi and W. T. Ford, *J. Am. Chem. Soc.*, **103**, 821 (1981); J. M. Fachet, P. Darling, and M. J. Farrall, *J. Org. Chem.*, **46**, 728 (1981); M. Bernard, W. T. Ford, T. W. Taylor, *Macromolecules*, **17**, 812 (1984); J. W. Goodwin, J. Hearn, C. Ho, and R. H. Ottewill, *Brit. Polym. J.*, **5**, 347 (1973).
17. T. Takeda and K. Kono, *Jpn. Kokai Tokkyo Koho* JP 63/66209A2 (3/24/88); *Chem. Abstr.* **109**, 150296j (1988); T. Kosugi and N. Nakayama, *Jpn. Kokai Tokkyo Koho* JP 62/235301 A2 (10/15/87); *Chem. Abstr.* **108**, 151155d (1988); H. A. Wright, U.S. Patent 4,500,692 (2/19/85); *Chem. Abstr.* **102**, 67706w (1985).
18. Y. Hashiguchi, M. Kishi, and T. Yugyu, *European Patent Appl.* EP 271922 A2 (6/22/88); *Chem. Abstr.* **109**, 129865j (1988).
19. N. Mitrea, M. Stanesca, C. Casadjicov, and A. Grigorescu, Rom. Patent Ro 92624 (9/30/87); *Chem. Abstr.* **109**, 93829j (1988).
20. F. Roger, M. Roustan, and H. Roques, *Entropic* **20** (120), 63 (1984); *Chem. Abstr.* **103**, 37785f (1985); S. M. Ahmed, *J. Dispersion Sci. Technol.*, **5** (3-4), 421 (1984).
21. H. A. Wright, U.S. Patent 4,500,692 (2/19/85).
22. M. Tanaka and T. Murishima, *Kagiku Kogaku Ronbutshu* **12** (2), 231 (1986).
23. K. Kasai, M. Hattori, O. Kikuchi, H. Takenchi, H. Hirai, and N. Sakurai, Eur. Patent Applic. EP 190886 A2 (8/13/86); *Chem. Abstr.* **105**, 191819t (1986).
24. Y. Hatate, A. Ikari, K. Kondo, and F. Nakashio, *Chem. Eng. Commun.* **34** (1-6), 325 (1985); *Chem. Abstr.* **103**, 22994z (1985).
25. A. S. Dunn, *Chem. Ind.* (*London*) p. 1406 (1971).
26. F. A. Bovey, I. M. Kolthoff, A. F. Medalia, and E. J. Meehan, "Emulsion Polymerization," Wiley (Interscience), New York, 1955; S. N. Sautin, P. A. Kulle, and N. I. Smirnov, *Zh. Prikl Khim* (*Leningrad*) **44**, 1569 (1971).
27. A. W. De Graff, "Continuous Emulsion Polymerization of Styrene in a One Stirred Tank Reactor," Lehigh Univ. Press, Bethlehem, Pennsylvania, 1970; O. Gellner, *Chem. Eng.* (*New York*) **73**, 74 (1966); A. G. Parts, D. E. Moore, and J. G. Waterson, *Makromol. Chem.* **89**, 156 (1965); E. W. Duck, *Encycl. Polym. Sci. Technol.* **5**, 801 (1966).
28. B. K. Dietrich, W. A. Pryor, and S. J. Wu, *J. Applied Polymer Sci.* **36** (5), 1129 (1988); *Chem. Abstr.* **109**, 93702n (1988).
29. I. M. Kolthoff and W. J. Dale, *J. Amer. Chem. Soc.* **69**, 441 (1947).
30. I. D. Robb, *J. Polym. Sci.*, *Part A-1* **7**, 417 (1969).

31. W. Schlenk, J. Appenrodt, A. Michael, and A. Thal, *Ber. Deut. Chem. Ges.* **47**, 473 (1914).
32. M. Szwarc, M. Levy, and R. Milkovich, *J. Amer. Chem. Soc.* **78**, 2656 (1956); M. Szwarc, "Carbanions Living Polymers and Electron Transfer Processes," Wiley (Interscience), New York, 1968.
33. D. J. Worsfold and S. Bywater, *J. Phys. Chem.* **70**, 162 (1966); F. S. Dainton *et al.*, *Makromol. Chem.* **89**, 257 (1965).
34. D. B. Priddy, U.S. Patent 4,647,632 (1987); D. B. Priddy and M. Piro, U.S. Patent 4,572,819 (1986).
35. D. B. Priddy, T. D. Traugott, and R. H. Seiss, *ACS Polymer Preprints* **30** (2), Sept. 1989, p. 195. Poster presentation at ACS national meeting, Sept. 1989, Miami Beach, Florida.
36. C. Harries, *Justus Liebigs Ann. Chem.* **383**, 213 (1911).
37. W. Schlenk, J. Appenrodt, A. Michael, and A. Thal, *Ber. Deut. Chem. Ges.* **47**, 473 (1914).
38. K. Ziegler and K. Bähr, *Ber. Deut. Chem. Ges.* **61**, 253 (1928).
39. W. Schlenk and E. Bergmann, *Justus Liebigs Ann. Chem.* **479**, 42 (1930).
40. K. Ziegler, F. Crossman, H. Kliener, and O. Schafter, *Justus Liebigs Ann. Chem.* **473**, 1 (1929).
41. L. M. Ellis, U.S. Patent 2,212,155 (1940).
42. W. C. E. Higginson and N. S. Wooding, *J. Chem. Soc., London*, p. 760 (1952).
43. M. Szwarc, *Nature* **178**, 1168 (1956); J. Smid and M. Szwarc, *J. Polym. Sci.* **61**, 31 (1962); M. Szwarc, "Carbanions, Living Polymers and Electron Transfer Processes," Wiley (Interscience), New York, 1968; M. Szwarc, *Encycl. Polym. Sci. Technol.* **8**, 303 (1968); T. Shimomura, J. Smid, and M. Szwarc, *J. Amer. Chem. Soc.* **89**, 5743 (1967).
44. D. J. Cram, *Chem. Eng. News* **41**, 92 (1963).
45. R. Woack and M. Doran, *Polymer* **2**, 365 (1961); D. J. Cram, *Chem. Eng. News* **41**, 92 (1963); A. A. Morton and E. Grovenstein, Jr., *J. Amer. Chem. Soc.* **74**, 5434 (1952); K. Yoshida and T. Morikawa, *Sci. Ind. Osaka* **27**, 80 (1953); W. E. Goode, W. H. Snyder, and R. C. Fettes, *J. Polym. Sci.* **42**, 367 (1960); W. H. Puterbaugh and C. R. Hauser, *J. Org. Chem.* **24**, 416 (1969).
46. R. Waack and M. A. Doran, *J. Org. Chem.* **32**, 3395 (1967).
47. A. A. Morton and E. Grovenstein, Jr., *J. Amer. Chem. Soc.* **74**, 5434 (1952).
48. A. A. Morton and E. J. Lanpher, *J. Polym. Sci.* **44**, 239 (1960).
49. M. Szwarc, *Encycl. Polym. Sci. Technol.* **8**, 303 (1968); "Carbanions, Living Polymers and Electron Transfer Processes," Wiley (Interscience), New York, 1968.
50. A. Noshay and J. McGrath, "Block Copolymers: Overview and Critical Survey," Academic Press, Orlando, Florida 1977; M. J. Folkes, "Processing, Structure and Properties of Block Copolymers," Elsevier Applied Sciences, Publishers, Ltd., Barking, UK, 1985; M. Morton, "Anionic Polymerization Principles and Practice," Academic Press, Orlando, Florida 1983; L. J. Fetters, *J. Polym. Sci., Polym. Sym.* **26**, 1 (1969).
51. D. R. Iyengar and T. J. McCarthy *in ACS Polymer Preprints*, **30** (2), Sept. 1989, p. 154. Paper presented at the ACS meeting, Sept. 1989, Miami Beach, Florida.

52. G. Odian, "Principles of Polymerization," 2nd ed., Wiley, New York, 1981.
53. M. Szwarc, M. Levy, and R. Milkovich, *J. Amer. Chem. Soc.* **78**, 2656 (1956).
54. Y. L. Spirin, A. A. Arest-Yakubovich, D. K. Polyakov, A. R. Gantmakher, and S. S. Medredev, *J. Polym. Sci.* **58**, 1181 (1962); D. N. Bhattacharyya, J. Smid, and M. Szwarc, *J. Phys. Chem.* **69**, 624 (1965).
55. C. G. Overberger, J. E. Mulvaney, and A. M. Schiller, *Encycl. Polym. Sci. Technol.* **2**, 95 (1965); M. Morton and L. J. Fetters, *Macromol. Rev.* **3**, 71 (1967).
56. B. Thomson, A. Rudin, and G. Lajoie, *J. Polym. Sci., Part A, Polym. Chem.* **33**, 345 (1995).
57. P. H. Plesch, ed., "Cationic Polymerization and Related Complexes," Academic Press, New York, 1954; "The Chemistry of Cationic Polymerization," Pergamon Press, New York, 1964.
58. F. S. Dainton and G. B. B. M. Sutherland, *J. Polym. Sci.* **4**, 37 (1949).
59. A. G. Evans, B. Holden, P. H. Plesch, M. Polanyi, H. A. Skinner, and M. A. Weinberger, *Nature (London)* **157**, 102 (1946); A. G. Evans, G. W. Meadows, and M. Polanyi, *ibid.* **158**, 94 (1946).
60. D. C. Pepper, *Quart. Rev., Chem. Soc.* **8**, 88 (1954).
61. D. C. Pepper, in "Friedel-Crafts and Related Reactions" (G. A. Olah, ed.), Vol. II, p. 123, Wiley (Interscience), New York, 1964.
62. A. G. Evans and G. W. Meadows, *J. Polym. Sci.* **4**, 359 (1949).
63. R. G. Heiligmann, *J. Polym. Sci.* **6**, 155 (1950).
64. J. A. Bittles, A. K. Chandhuri, and S. W. Benson, *J. Polym. Sci., Part A* **2**, 1221 (1964); G. F. Endres and C. G. Overberger, *J. Amer. Chem. Soc.* **77**, 2201 (1955); D. O. Jordan and A. R. Mathieson, *J. Chem. Soc., London* p. 611 (1952); J. A. Bittles, A. K. Chandhuri, and S. W. Benson, *J. Polym. Sci., Part A* **2**, 1221 (1964).
65. F. C. Whitmore, *Ind. Eng. Chem.* **26**, 94 (1964).
66. R. Simha and L. A. Wall, in "Catalysis" (P. H. Emmett, ed.), Vol. VI, p. 266, Van Nostrand-Reinhold, Princeton, New Jersey, 1958.
67. J. Hine, "Physical Organic Chemistry," p. 219, McGraw-Hill, New York, 1956.
68. Y. Tsuda, *Macromol. Chem.* **36**, 102 (1960).
69. J. P. Kennedy and E. Maechal, "Carbocationic Polymerization," Wiley, New York, 1981.
70. G. A. Olah, ed., "Friedel-Crafts and Related Reactions," Vol. I, Wiley (Interscience), New York, 1963.
71. D. N. P. Satchell, *J. Chem. Soc., London* pp. 1453 and 3822 (1961).
72. A. M. Eastham, *Encycl. Polym. Sci. Technol.* **3**, 35 (1965).
73. J. P. Kennedy, *J. Org. Chem.* **35**, 532 (1970).
74. J. P. Kennedy, in "Polymer Chemistry of Synthetic Elastomers" (J. P. Kennedy and E. Tarnquist, eds.), Part 1, Chapter 5A, p. 291, Wiley (Interscience), New York, 1968; Belgian Patent 663,319 (1965).
75. C. E. Schildknecht, "Vinyl and Related Polymers," Chapter 10, Wiley, New York, 1952.
76. R. M. Thomas, W. J. Sparks, and P. K. Frolich, *J. Amer. Chem. Soc.* **62**, 276 (1940).
77. R. M. Thomas and W. J. Sparks, U.S. Patent 2,356,128 (1944).

78. J. D. Calfee and R. M. Thomas, U.S. Patent 2,431,461 (1947).
79. J. P. Kennedy and R. G. Squires, *Polymer* **6**, 579 (1965).
80. S. G. Gallo, H. K. Wiese, and J. F. Nelson, *Ind. Eng. Chem.* **40**, 1277 (1948).
81. J. P. Kennedy, U.S. Patent 3,349,065 (1967).
82. C. S. Marvel, R. Gilkey, C. R. Morgan, J. F. Noth, R. D. Rands, Jr., and H. C. Young, *J. Polym. Sci.* **6**, 483 (1951).
83. A. A. Morton, M. L. Brown, and E. Magat, *J. Amer. Chem. Soc.* **68**, 161 (1946).
84. C. E. Schildknecht, A. O. Zoss, and C. McKinley, *Ind. Eng. Chem.* **39**, 180 (1947).
85. M. Fischer, German Patent 874,215 (1953).
86. D. F. Herman and W. K. Nelson, *J. Amer. Chem. Soc.* **75**, 3877 (1953).
87. K. Ziegler, Belgian Patent 533,362 (1954); K. Ziegler, E. Holzkamp, H. Breil, and H. Martin, *Angew. Chem.* **67**, 541 (1955).
88. G. Natta, *Atti Accad. Naz. Lincei, Cl. Sci. Fis., Mat. Natur., Rend.* [8] **4**, 61 (1955); G. Natta, P. Pino, P. Corradini, F. Danusso, E. Mantica, G. Mazzanti, and G. Moraglio, *J. Amer. Chem. Soc.* **77**, 1708 (1955); G. Natta, L. Porri, and S. Valenti, *Makromol. Chem.* **67**, 225 (1963).
89. L. Porri, G. Natta, and M. C. Gallazzi, *J. Polym. Sci., Part C* **16**, 2525 (1967); V. A. Kormer, B. D. Babitskii, M. I. Lobach, and N. N. Chesnokova, *ibid.*, p. 4351 (1969).
90. G. Wilke, *Angew Chem.* **75**, 10 (1963); Y. Tajima and E. Kunioka, *J. Polym. Sci., Part B* **5**, 221 (1967); G. Wilke, M. Kroener, and B. Bogdanovic, *Angew. Chem.* **73**, 755 (1961); B. D. Babitski, V. A. Kormer, I. M. Lapak, and V. I. Soblikova, *J. Polym. Sci., Part C* **16**, 3219 (1968).
91. W. Cooper, *Ind. Eng. Chem., Prod. Res. Develop.* **9**, 457 (1970).
92. W. Cooper and G. Vaughn, *Progr. Polym. Sci.* **7** (1967); J. Boor, *Macromol Rev.* **2**, 115 (1967); J. P. Kennedy and A. W. Langer, Jr., *Fortsch. Hochpolym.-Forsch.* **3**, 539 (1964); N. G. Gaylord and H. Mark, "Linear and Stereoregular Addition Polymers," Wiley (Interscience), New York, 1959; K. Ziegler, *Angew. Chem.* **76**, 545 (1964); J. P. Kennedy and G. E. Milliman, Jr., *Advan. Chem. Ser.* **91**, 287–305 (1969); V. A. Kormer, B. D. Babitskiy, and M. I. Lobach, *ibid.*, pp. 306–316; G. Natta and U. Giannini, *Encycl. Polym. Sci. Technol.* **4**, 137 (1966).
93. L. R. Gilliom and R. H. Grubbs, *J. Amer. Chem. Soc.* **108**, 733 (1986); E. V. Anslyn and R. H. Grubbs, *J. Amer. Chem. Soc.* **109**, 4880 (1987).
94. F. N. Tebbe, G. W. Parshall and G. S. Reddy, *J. Amer. Chem. Soc.* **100**, 3611 (1978).
95. L. R. Gilliom and R. H. Grubbs, *J. Amer. Chem. Soc.* **108**, 733 (1986).
96. *Chemical Marketing Reporter*, Dec. 23, 1996, pp. 5 and 18.
97. D. D. Devore, L. H. Crawford, J. C. Stevens, F. J. Timmers, R. D. Mussell, D. R. Wilson, and R. K. Rosen, World Patent Appl. WO 9500526 (1/5/95); *Chem. Abstr.* **123**, 257778 (1995).
98. *Chem. Eng. News*, Jan. 20, 1997, p. 14.
99. J. H. Canich and B. J. Folie, U.S. Patent 5,408,017 (4/18/95). To Exxon.
100. Patents and publications by Prof. Tobin Marks at Northwestern University. For example, see T. J. Marks and Xyang, U.S. Patent 5,447,895 (9/5/95).

2
POLYESTERS

Polyesters are polymers with repeating carboxylate groups

$$-\underset{\underset{O}{\|}}{C}O-$$

in their backbone chain. Polycarbonates are esters of carbonic acid and will be discussed in Chapter 3.

Polyesters are synthesized [1-12] by the typical esterification reactions, which can be generalized by the reaction shown in Eq. (1):

$$R\overset{O}{\overset{\|}{C}}-X + N: \rightleftharpoons \left[R-\underset{\underset{X}{|}}{\overset{\overset{\ddot{O}}{|}}{C}}-N \right] \longrightarrow R\overset{O}{\overset{\|}{C}}-N + X: \quad (1)$$

where N: is a nucleophilic reagent such as $\overline{O}R'$. The rate of reaction will be dependent on the structure of R, R', X, N and on whether a catalyst is used.

Tartaric acid-glycerol polyesters were reported in 1847 by Berzelius [13] and those of ethylene glycol and succinic acid were reported by Lorenzo in 1863 [14]. Carothers extended much of the earlier work and helped to clarify the understanding of the polyesterification reaction in light of the knowledge of polymer chemistry at that time. Polyethylene terephthalate [15] and the polyadipates [16] (for polyurethane resins) were the first major commercial applications of polyesters.

The major synthetic methods used to prepare polyesters all involve condensation reactions as shown in Eqs. (2)–(12):

$$\text{HOR—COOH} \xrightarrow{[17]} \text{H}\!\left[\text{OR}-\overset{\text{O}}{\underset{\|}{\text{C}}}\right]_n\!\text{OH} \tag{2}$$

$$\text{R'COOR—COOH} \xrightarrow{[18]} \text{R'CO}\!\left[\text{OR}\overset{\text{O}}{\underset{\|}{\text{C}}}\right]_n\!\text{OH} \tag{3}$$

$$\text{RCOOROCOR} + \text{R(COOH)}_2 \xrightarrow{[19]} \text{RCO}\!\left[\text{OR}\overset{\text{O}}{\underset{\|}{\text{C}}}\right]_n\!\text{OH} \tag{4}$$

$$\text{HO—R—OH} + \text{R'(COOH)}_2 \xrightarrow{[19,\,20]} \text{H}\!\left[\text{OROC}\overset{\text{O}\ \ \text{O}}{\underset{\|\ \ \|}{\text{R'C}}}\right]_n\!\text{OH} \tag{5}$$

$$\text{HO—R—OH} + \text{R'(COOR'')}_2 \xrightarrow{[21,\,22]} \text{H}\!\left[\text{OR}-\text{OC}-\overset{\text{O}\ \ \text{O}}{\underset{\|\ \ \|}{\text{R'C}}}\right]_n\!\text{OH} \tag{6}$$

$$\text{HO—R—OH} + \text{R'(COX)}_2 \xrightarrow{[23,\,24]} \text{H}\!\left[\text{OROC}\overset{\text{O}\ \ \text{O}}{\underset{\|\ \ \|}{\text{R'C}}}\right]_n\!\text{X} \tag{7}$$

$$\text{NaOR—ONa} + \text{R'(COX)}_2 \xrightarrow{[25]} \text{Na}\!\left[\text{OROC}\overset{\text{O}\ \ \text{O}}{\underset{\|\ \ \|}{\text{R'C}}}\right]_n\!\text{X} \tag{8}$$

$$(\text{X} = \text{Cl},\ \text{OCH}_3)$$

$$\text{HO—R—OH} + \text{R'(CO)}_2\text{O} \xrightarrow{[26]} \text{H}\!\left[\text{OROC}\overset{\text{O}\ \ \text{O}}{\underset{\|\ \ \|}{\text{R'C}}}\right]_n\!\text{OH} \tag{9}$$

$$\text{HO—R—OH} + \text{R'O}\overset{\text{O}}{\underset{\|}{\text{C}}}\text{OR'} \xrightarrow{[27]} \text{H}\!\left[\text{OROC}\overset{\text{O}}{\underset{\|}{}}\right]_n\!\text{OR'} \tag{10}$$

$$Br-R-Br + R'(COOAg)_2 \xrightarrow{[28]} Br\left[OROCR'C\underset{\underset{O}{\|}}{\overset{\overset{O}{\|}}{}}\right]_n Ag \quad (11)$$

$$\left[\begin{array}{c}-(CH_2)_n-\overset{\overset{O}{\|}}{C}-\\-O-\end{array}\right] \xrightarrow{[29]} \left[O(CH_2)_n-\overset{\overset{O}{\|}}{C}\right]_n \quad (12)$$

The use of triols or tricarboxylic acids leads to crosslinked or network polyesters. For example, an alkyd resin is formed by the reaction of glycerol with phthalic anhydride [30].

Some typical examples of the preparation and properties of some representative polyesters are shown in Table I.

2-1. Preparation of Poly(ethylene terephthalate) [31]

$$CH_3O\overset{\overset{O}{\|}}{C}-\bigcirc-\overset{\overset{O}{\|}}{C}-OCH_3 + 2HOCH_2CH_2OH \xrightarrow{-2CH_3OH}$$

$$HOCH_2CH_2O\overset{\overset{O}{\|}}{C}-C_6H_4-\overset{\overset{O}{\|}}{C}-OCH_2CH_2OH$$

$$\underset{\text{heat}}{\Big\downarrow} {-HOCH_2CH_2OH}$$

$$\left[OCH_2CH_2O\overset{\overset{O}{\|}}{C}-C_6H_4-\overset{\overset{O}{\|}}{C}\right]_n \quad (13)$$

■ **CAUTION:** This reaction should be carried out in a hood behind a protective shield.

To a weighed thick-walled glass tube with a constricted upper portion for vacuum tube connection and equipped with a metal protective sleeve is added 15.5 gm (0.08 mole) of dimethyl terephthalate (see Note a), 11.8 gm (0.19 mole) of ethylene glycol (Note b), 0.025 gm of calcium acetate dihydrate, and 0.006 gm of antimony trioxide. The tube is warmed gently in an oil bath to melt the mixture and then a capillary tube connected to a nitrogen source is placed in the melt. While heating to 197°C a slow stream of nitrogen is passed through the melt to help eliminate the methanol.

TABLE I

PREPARATION OF POLYESTERS BY CONDENSATION REACTIONS

Reactants		Catalyst (gm)	Reaction conditions		m.p. (°C)	Mol. wt.	Ref.
Alcohol	Diacid or derivative		Temperature (°C)	Time (hr)			
$HO(CH_2)_nOH$ $n = 2,3,6,10$	Aliphatic type: carbonic, oxalic, succinic, glutaric, adipic, pimelic, and sebacic acids	—	—	—	—	—	a
$HO(CH_2)_nOH$ $n = 2,3,6,10$	Phthalic acid	—	—	—	—	—	a
$HO(CH_2)_2OH$	Isophthaloyl chloride	—	40–218	$\frac{1}{2}-\frac{3}{4}$	140–142	—	b
$HO(CH_2)_nOH$ $n = 2,3,6,8,10,18$	Terephthalic acid	—	—	—	—	—	c
$HOCH_2(CF_2)_3CH_2OH$	$(CH_2)_4(COOH)_2$	$ZnCl_2$ (0.01)	150–215	240	Visc. liq.	4000	d
$HOCH_2(CF_2)_3CH_2OH$	$(CF_2)_4(COCl)_2$	—	—	21.5	Approx. 35	M_n, 6570	d
$HO(CH_2)_{10}OH$	1,1-Dichlorocarbonylferrocene	—	—	—	25–65	500–4000	e
$HO(CH_2)_4OH$	Dimethyl sebacate	PbO (0.0625)	183–259	3	64–65	—	f
$HOCH_2HC=CH-CH_2OH$	Dimethyl sebacate	PbO (0.0625)	100–142 (2 mm Hg)	18–24	49	—	f

[a] W. H. Carothers, U.S. Patent 2,071,053 (1937).
[b] P. J. Flory and F. S. Leutner, U.S. Patent 2,623,034 (1952).
[c] J. R. Whinfield and J. T. Dickson, British Patent 578,079 (1946); E. F. Izard, *J. Polym. Sci.* **8**, 503 (1952); J. R. Caldwell and R. Gilkey, U.S. Patent 2,891,930 (1959).
[d] G. C. Schweiker and P. Robitschek, *J. Polym. Sci.* **24**, 33 (1957).
[e] C. U. Pittman, Jr., *J. Polym. Sci., Part A-1* **6**, 1687 (1968).
[f] C. S. Marvel and J. H. Johnson, *J. Amer. Chem. Soc.* **72**, 1674 (1950).

The tube is heated for 2–3 hr at 197°C, or until all the methanol has been removed (Note c). The side arm is also heated to prevent clogging by the condensation of some dimethyl terephthalate. The polymer tube is next heated to 222°C for 20 min and then at 283°C for 10 min. The side arm is connected to a vacuum pump and the pressure is reduced to 0.3 mm Hg or less while heating at 283°C for 3.5 hr. The tube is removed from the oil bath, cooled (Note d) behind a safety shield, and then weighed. The yield is quantitative if no loss of dimethyl terephthalate has occurred. The polymer melts at about 270°C. (The crystalline melting point is 260°C.) The inherent viscosity of a 0.5% solution in *sym*-tetrachloroethanol/phenol (40/60) is approx. 0.6–0.7 at 30°C.

NOTES: (a) Dimethyl terephthalate is the best grade or is recrystallized from ethanol, m.p. 141°–142°C. (b) The ethylene glycol is anhydrous reagent grade or prepared by adding 1 gm of sodium/100 ml, refluxing for 1 hr in a nitrogen atmosphere, and distilling, b.p. 196°–197°C. (c) Failure to remove all the methanol leads to low molecular weight polymers. (d) On cooling the polymer contracts from the walls and this may cause the tube to shatter.

Higher molecular weight polyethylene terephthalate can be prepared by heating in a solid state polymerization reactor at 230°C using a nitrogen gas stream. The intrinsic viscosity is raised in one case from 0.59 dl/gm to 0.90 dl/gm after 5 hr (nitrogen flow = 4 standard cubic feet/hr) [32].

Block or graft polyesters of polyethylene terephthalate can also be prepared by heating in the molten state with 0.5 to 1.5% phosphites (such as triphenylphosphite) [33] for 2.5–12.5 hr. The intrinsic viscosity [η] increased only in some of the cases (0.5% concentration of polymer in a 60:40 phenol-tetrachloroethane solvent). The polyesters thus produced have increased tensile strength, durability, and impact resistance.

REFERENCES

1. J. M. Hawthorne and C. J. Heffelfinger, *Encycl. Polym. Sci. Technol.* **11**, 1 (1969).
2. I. Goodman, *Encycl. Polym. Sci. Eng.* **12**, 1 (1988).
3. I. Goodman, *Encycl. Polym. Sci. Technol.* **11**, 62 (1969).
4. H. V. Boening, *Encycl. Polym. Sci. Technol.* **11**, 729 (1969).

5. V. V. Korshak and S. V. Vinogradova, "Polyesters," Elsevier, Amsterdam, 1953.
6. V. V. Korshak and S. V. Vinogradova, "Polyesters," Pergamon, Oxford, 1969.
7. R. Hill and E. E. Walker, *J. Polym. Sci.* **3**, 609 (1948).
8. H. J. Hagemeyer, U.S. Patent 3,043,808 (1962).
9. H. Batzer and F. Wiloth, *Makromol. Chem.* **8**, 41 (1952).
10. B. M. Grievson, *Polymer* **1**, 499 (1960).
11. M. J. Hurwitz and E. W. Miller, French Patent 1,457,711 (1966).
12. Borg-Warner Corp., British Patent 1,034,194 (1966).
13. J. Berzelius, *Rapp. Annu.* **26**, 1 (1847); *Jahresbericht* **12**, 63 (1833).
14. A. V. Lorenzo, *Ann. Chim. Phys.* [2] **67**, 293 (1863).
15. J. R. Whinfield, *Nature (London)* **158**, 930 (1946); S. K. Agarawal, *Indian Chem. J.* **5**, 25 (1970).
16. O. Bayer, *Justus Liebigs Ann. Chem.* **549**, 286 (1941).
17. W. H. Carothers and E. J. Van Natta, *J. Amer. Chem. Soc.* **55**, 4714 (1933).
18. R. Gilkey and J. R. Caldwell, *J. Appl. Polym. Sci.* **2**, 198 (1959).
19. E. R. Walsgrove and F. Reeder, British Patent 636,429 (1950).
20. R. E. Wilfong, *J. Polym. Sci.* **54**, 385 (1961).
21. C. S. Marvel and J. H. Johnson, *J. Amer. Chem. Soc.* **72**, 1674 (1950).
22. J. T. Dickson, H. P. W. Huggill, and J. C. Welch, British Patent 590,451 (1947).
23. P. J. Flory and F. S. Leuther, U.S. Patents 2,623,034 and 2,589,688 (1952).
24. K. Yamaguchi, M. Takayanagi, and S. Kuriyama, *J. Chem. Soc. Jap. Ind. Chem. Sect.* **58**, 358 (1955).
25. J. T. Dickson, H. P. W. Huggill, and J. C. Welch, British Patent 590,451 (1947).
26. R. H. Kienle and H. G. Hovey, *J. Amer. Chem. Soc.* **52**, 3636 (1930); W. H. Carothers and J. W. Hill, *ibid.* **54**, 1577 (1932).
27. C. A. Bishoff and A. von Hendenström, *Chem. Ber.* **35**, 3431 (1902).
28. W. H. Carothers and J. A. Arvin, *J. Amer. Chem. Soc.* **51**, 2560 (1929); D. Vorländer, *Justus Liebigs Ann. Chem.* **280**, 167 (1894).
29. F. E. Critchfield and R. D. Lundberg, French Patent Appl. 2,026,274 (1970).
30. M. Callahan, U.S. Patents 1,191,732 and 1,108,329-39 (1914).
31. J. R. Whinfield and J. T. Dickson, British Patent 578,079 (1946); J. R. Whinfield, *Nature (London)* **158**, 930 (1946).
32. P. R. Wendling, U.S. Patent 4,532,319 (1985).
33. S. M. Aharoni and D. Masolamani, U.S. Patent 4,568,720 (1986).

3
POLYCARBONATES

Structurally, polycarbonates are polyesters derived from the reaction of carbonic acid or its derivatives with dihydroxy compounds (aliphatic, aromatic, or mixed type compounds).

$$H\left[OROC\overset{O}{\underset{\|}{}}\right]_n OROH$$

Polycarbonates are almost unaffected by water and many inorganic and organic solvents. These properties allow these resins to fill many applications, especially in view of their attractive mechanical properties. The stability of polycarbonates is good below 250°C and this property has been reviewed [1].

Einhorn [2] first prepared polycarbonates by the Schotten–Baumann reaction of phosgene with either hydroquinone or resorcinol in pyridine (Eq. 1). Bischoff and von Hedenström [3] later reported that the same polycarbonates can be prepared from the ester-interchange reaction of hydroquinone or resorcinol with diphenyl carbonate (Eq. 1).

$$n\text{HO}-\text{R}-\text{OH} + n\text{COCl}_2 \xrightarrow{-2n\text{HCl}} \left[\text{OR}-\text{O}\overset{O}{\underset{\|}{\text{C}}}\right]_n \quad (1)$$

$$n\text{HO}-\text{R}-\text{OH} + n\text{C}_6\text{H}_5\text{O}-\overset{O}{\underset{\|}{\text{C}}}-\text{OC}_6\text{H}_5 \xrightarrow{-2n\text{C}_6\text{H}_5\text{OH}}$$

Carothers and Hill [4] were first to systematically study methods to prepare high molecular weight polymers. Later Peterson [5] improved the procedure and developed a way to prepare high molecular weight polycarbonates having useful film and fiber properties.

The subject of polycarbonates has been reviewed earlier and many references to their applications are given [6-9].

Polycarbonates of 2,2-bis(4-hydroxyphenyl) propane (bisphenol A) are commercially the most important polycarbonates [6, 7]. Polycarbonates are also prepared by the reaction of bischloroformates with dihydroxy compounds.

Aliphatic polycarbonates are also receiving some attention but are less important commercially.

General Electric announced [10] the development of technology leading to cyclic oligomers that can be converted to polycarbonates using anionic catalysts. The cyclic oligomer with up to 240 rings can be prepared from bisphenol A bischloroformate using triethylamine and sodium hydroxide as described later in this chapter.

U.S. consumption of polycarbonates in 1996 amounted to about 880 million lbs and is estimated to grow to 1.3 billion lbs in 2001. The major domestic producers are Dow, GE, and Bayer AG. The major end uses include windows (20%), transportation (25%), electrical/electronic devices (24%), industrial uses (8%), and other uses account for 23%.

3-1. Preparation of Poly(2,2-dimethyl-1,3-propanediol carbonate) [poly(neopentylene carbonate)] [11]

$$(C_2H_5O)_2O + HOCH_2-\underset{\underset{CH_3}{|}}{\overset{\overset{CH_3}{|}}{C}}-CH_2OH \longrightarrow \left[CH_2-\underset{\underset{CH_3}{|}}{\overset{\overset{CH_3}{|}}{C}}-CH_2OCOO \right]_{20-22}$$

$$+ 2C_2H_5OH \quad (2)$$

To a flask equipped with a 3 ft Vigreaux column is added 20.8 gm (0.2 mole) of neopentyl glycol, 24 gm (0.2 mole) of diethyl carbonate, and 0.05 gm of dry sodium methoxide. The mixture is gradually heated and the ethanol that is formed is distilled off at 77°-84°C. After the ethanol stops distilling the pressure is reduced and the mixture heated to 200°C to complete the reaction. The residue

(25 gm) on cooling is an opaque, glassy, and tough polymer. The polymer is purified from ligroin–benzene to afford 23.5 gm (90%) of a white powder, m.p. 107°–109°C. Intrinsic viscosity of the polymer is 0.035; mol. wt., 2500. The polymer is soluble in benzene but insoluble in most other organic solvents. The polymer is stable to 1 N aq. ethanolic NaOH on refluxing for 2–3 hr.

REFERENCES

1. A. Davis and J. H. Golden, *J. Macromol. Sci. Rev. Macromol. Chem.* **3**, 49 (1969).
2. A. Einhorn, *Justus Liebigs Ann. Chem.* **300**, 135 (1890).
3. C. A. Bischoff and A. von Hedenström, *Chem. Ber.* **35**, 3431 (1902).
4. W. H. Carothers and J. W. Hill, *J. Amer. Chem. Soc.* **54**, 1559, 1566, and 1579 (1932).
5. W. R. Peterson, U.S. Patent 2,210,817 (1940).
6. H. Schnell, *Angew. Chem.* **68**, 633 (1956).
7. H. Schnell, "Chemistry and Physics of Polycarbonates," Wiley (Interscience), New York, 1964.
8. L. Bottenbruch, *Encycl. Polym. Sci. Technol.* **10**, 710 (1969).
9. W. F. Christopher and D. W. Fox, "Polycarbonates," Van Nostrand-Reinhold, Princeton, New Jersey, 1962.
10. S. Stinson, *Chem. Eng. News*, Sept. 18, 1989, p. 8.
11. S. Sarel and L. A. Pohoryles, *J. Amer. Chem. Soc.* **80**, 4596 (1958).

4
POLYAMIDES

Polymeric amides are found in nature in the many polypeptides (proteins) which constitute a variety of animal organisms and the composition of silk and wool. It was a search to prepare substitutes for the latter which led to the commercial development of the synthetic polyamides known as nylons.

Several early investigators reported polyamide type materials. Balbiano and Trasciatti [1] heated a mixture of glycine in glycerol to give a yellow amorphous glycine polymer [2]. Manasse [3] obtained nylon 7 by heating 7-aminoheptanoic acid to the melting point (Eq. 1). Curtius [4] found that ethyl glycinate

$$n\text{NH}_2(\text{CH}_2)_6\text{COOH} \longrightarrow \text{\textlbrackdbl}\text{NH}(\text{CH}_2)_6\text{CO}\text{\textrbrackdbl}_n + n\text{H}_2\text{O} \quad (1)$$

polymerizes on standing in the presence of moisture or in the dry state in ethyl ether to afford polyglycine [5].

$$n\text{NH}_2\text{CH}_2\text{COOC}_2\text{H}_5 \longrightarrow \text{\textlbrackdbl}\text{NHCH}_2\text{CO}\text{\textrbrackdbl}_n + n\text{C}_2\text{H}_5\text{OH} \quad (2)$$

The β-amino acids on heating afford ammonia but no polymers. In addition, the γ- and δ-amino acids on heating afford stable lactams and no polymers. However, the ϵ-, ζ-, and η-amino acids give polyamides.

The reaction of dibasic acids with diamines was reported in the early literature [6–13] to give low molecular weight cyclic amides as infusible and insoluble products. It was Carothers [14–18] who first recognized that polymeric amides were formed by the reaction of diamines with dibasic acids. Many of the polyamides were able to be spun into fibers and the fibers were called "Nylon" by Du Pont [14–20].

The major methods of preparing polyamides are summarized in Scheme 1. Other methods of less importance are summarized in Scheme 2.

Additional developments in polyamide synthesis involve the preparation of Nylon-1 by the anionic polymerization of alkylisocyanates at temperatures below $-20°C$ [21] (Eq. 2a).

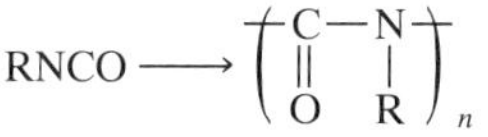

$$\text{RNCO} \longrightarrow \left(\begin{array}{cc} \text{C} - \text{N} \\ \| \quad | \\ \text{O} \quad \text{R} \end{array} \right)_n \tag{2a}$$

Nylon 4,6 or poly(tetramethylenediamine-co-adipic acid) melts approx. 30°C higher than Nylon 6,6 (i.e., 295°C vs 265°C). There are

SCHEME 1
THE MAJOR METHODS FOR THE SYNTHESIS OF POLYAMIDES

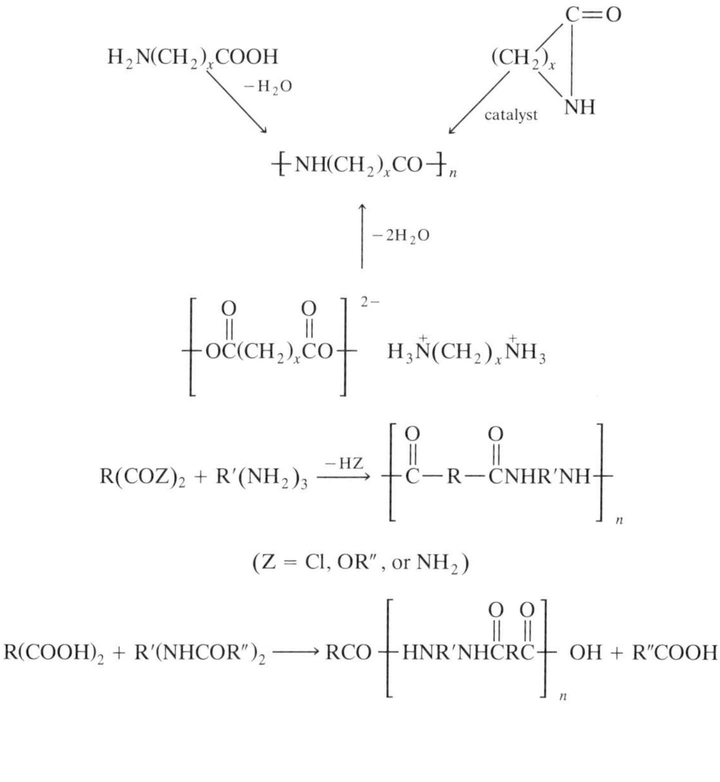

$$H_2N(CH_2)_xCOOH \xrightarrow{-H_2O} \quad (CH_2)_x \xrightarrow[\text{catalyst}]{C=O} \xrightarrow{NH}$$

$$\dashv NH(CH_2)_xCO \vdash_n$$

$$\uparrow -2H_2O$$

$$\left[\dashv OC(CH_2)_xCO \vdash \right]^{2-} \quad H_3\overset{+}{N}(CH_2)_x\overset{+}{N}H_3$$

$$R(COZ)_2 + R'(NH_2)_2 \xrightarrow{-HZ} \left[\dashv \overset{O}{\underset{\|}{C}} - R - \overset{O}{\underset{\|}{C}}NHR'NH \vdash \right]_n$$

$(Z = Cl, OR'', \text{ or } NH_2)$

$$R(COOH)_2 + R'(NHCOR'')_2 \longrightarrow RCO \left[HNR'NH\overset{O}{\underset{\|}{C}}R\overset{O}{\underset{\|}{C}} \right]_n OH + R''COOH$$

SCHEME 2
MISCELLANEOUS METHODS FOR THE SYNTHESIS OF POLYAMIDES FROM NITRILES

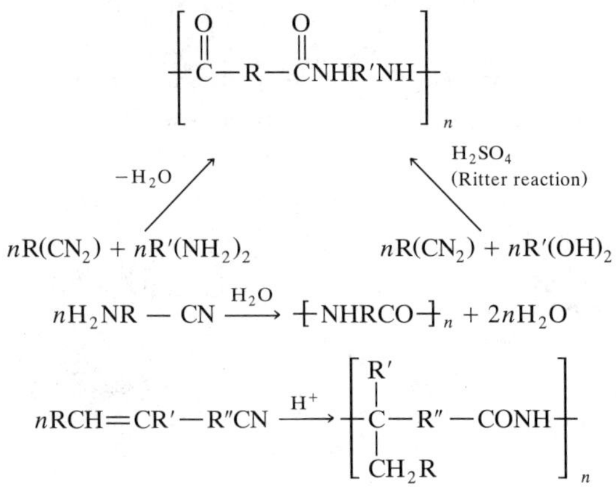

$$\left[\begin{matrix} O & O \\ \| & \| \\ -C-R-CNHR'NH- \end{matrix} \right]_n$$

$$nR(CN_2) + nR'(NH_2)_2 \xrightarrow{-H_2O} \quad \xleftarrow{H_2SO_4 \text{ (Ritter reaction)}} nR(CN_2) + nR'(OH)_2$$

$$nH_2NR-CN \xrightarrow{H_2O} \{NHRCO\}_n + 2nH_2O$$

$$nRCH=CR'-R''CN \xrightarrow{H^+} \left[\begin{matrix} R' \\ | \\ -C-R''-CONH- \\ | \\ CH_2R \end{matrix} \right]_n$$

processes reported to prepare it by a solid-phase polymerization to give polymers with $\overline{M}_n$ of about 33,000 [22]. The fibers can be spun at 305°–330°C to give filaments of high modulus that are used in tire cords [23].

The first commercial Aramid fiber was based on poly(*m*-phenylene isophthalamide) and has the tradename Nomex (Du Pont) [24].

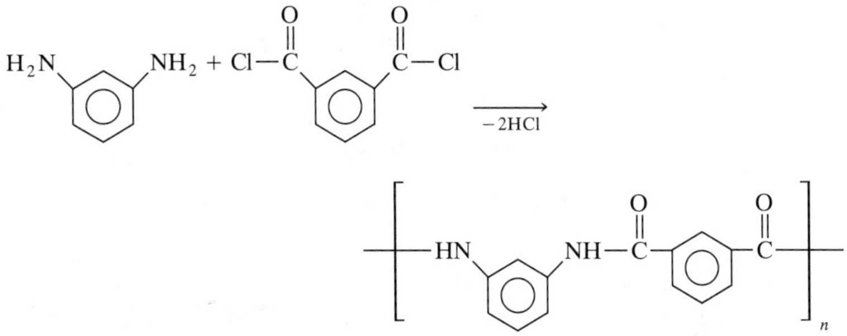

$$H_2N-\phi-NH_2 + Cl-\overset{O}{\underset{\|}{C}}-\phi-\overset{O}{\underset{\|}{C}}-Cl \xrightarrow{-2HCl}$$

$$\left[-HN-\phi-NH-\overset{O}{\underset{\|}{C}}-\phi-\overset{O}{\underset{\|}{C}}- \right]_n \quad (2b)$$

Du Pont also introduced Kevlar® based on poly(*p*-phenyleneterephthalamide) [25].

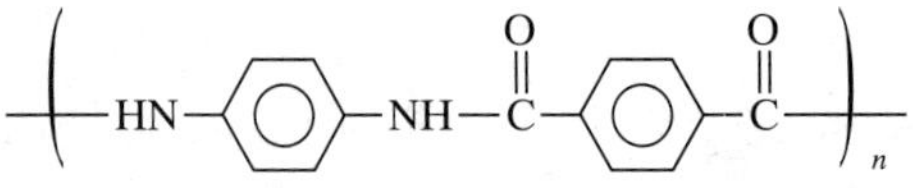

$$\left(-HN-\phi-NH-\overset{O}{\underset{\|}{C}}-\phi-\overset{O}{\underset{\|}{C}}- \right)_n$$

The Aramid fibers have high tensile strength and good flame resistance.

The synthesis of optically active polyamides, or Nylons, is a growing area of interest.

The preparation of polyamides has been reviewed and these reviews are worth consulting [26–29].

The use of the low-temperature interfacial condensation technique to prepare polyamides and various other polymers has recently been reviewed in a monograph on this subject [30, 31].

Some of the important variables involved in interfacial polymerization are (a) organic solvent, (b) reactant concentration, and (c) use of added detergents [32].

The organic solvent is the most important variable since it controls partition and diffusion of the reactants between the two immiscible phases, the reaction rate, solubility, and swelling of permeability of the growing polymer. The solvent should be of such composition as to prevent precipitation of the polymer before a high molecular weight has been attained. The final polymer does not have to dissolve in the solvent. The type of solvent will also influence the characteristics of the physical state of the final polymer. Solvents such as chlorinated or aromatic hydrocarbons make useful solvents in this system.

Concentrations in the range of approximately 5% polymer based on the combined weights of water and organic solvent usually are optimum. Concentrations that are too low may lead to hydrolysis of the acid halide and concentrations that are too high may cause excessive swelling of the solvent in the polymer.

In some cases the addition of 0.2–1% of sodium lauryl sulfate has been found to give satisfactory results. In many cases it may not be necessary.

The reactants should be pure but need not be distilled prior to use. An exact reactant balance is not as essential as in the melt polymerization process since the reactions are extremely rapid. A slight excess (5–10%) of the diamine usually helps to produce higher molecular weights.

The advantage of the interfacial polymerization process is that it is a low-temperature process requiring ordinary equipment. It also allows one to prepare those polyamides that are unstable in the melt polymerization process. Random or block polymers can easily be prepared depending on the reactivity of the reactants and their mixing (consecutively versus all at once).

4-1. Preparation of Poly(hexamethylenesebacamide) (Nylon 6-10) by the Interfacial Polymerization Technique [33]

$$(CH_2)_6(NH_2)_2 + (CH_2)_8(COCl)_2 \longrightarrow \left[HN-(CH_2)_6-NH\overset{O}{\underset{\|}{C}}-(CH_2)_8-\overset{O}{\underset{\|}{C}} \right]_n \quad (3)$$

To a tall-form beaker is added a solution of 3.0 ml (0.014 mole) of sebacoyl chloride dissolved in 100 ml of distilled tetrachloroethylene. Over this acid chloride solution is carefully poured a solution of 4.4 gm (0.038 mole) of hexamethylenediamine (see Note) dissolved in 50 ml of water. The polyamide film which begins to form at the interface of these two solutions is grasped with tweezers or a glass rod and slowly pulled out of the beaker in a continuous fashion. The process stops when one of the reactants becomes depleted. The resulting "rope"-like polymer is washed with 50% aqueous ethanol or acetone, dried, and weighed to afford 3.16–3.56 gm (80–90%) yields of polyamide, $\eta_{inh} = 0.4$ to 1.8 (m-cresol, 0.5% conc. at 25°C), m.p. 215°C (soluble in formic acid).

NOTE: In this experiment, an excess of diamine is used to act as an acid acceptor.

REFERENCES

1. L. Balbiano and D. Trasciatti, *Ber. Deut. Chem. Ges.* **33**, 2323 (1900).
2. L. C. Maillard, *Ann. Chim. Anal. Chim. Appl.* [9] **1**, 519 (1914); [9] **2**, 210 (1914).
3. A. Manasse, *Ber. Deut. Chem. Ges.* **35**, 1367 (1902).
4. I. Curtius, *Ber. Deut. Chem. Ges.* **37**, 1284 (1904).
5. M. Frankel and A. Katchalsky, *Nature* **144**, 330 (1939).
6. E. Fischer and H. Koch, *Justus Liebigs Ann. Chem.* **232**, 227 (1886).
7. A. W. Hoffman, *Ber. Deut. Chem. Ges.* **5**, 247 (1872).
8. M. Freund, *Ber. Deut. Chem. Ges.* **17**, 137 (1884).
9. F. Anderlini, *Gazz. Chim. Ital.* **24** (1) 397 (1894).
10. E. Fischer, *Ber. Deut. Chem. Ges.* **46**, 2504 (1913).
11. H. Meyer, *Justus Liebigs Ann. Chem.* **347**, 17 (1906).
12. P. Ruggli, *Justus Liebigs Ann. Chem.* **392**, 92 (1912).
13. C. L. Butler and R. Adams, *J. Amer. Chem. Soc.* **47**, 2614 (1925).
14. W. H. Carothers and G. J. Berchet, *J. Amer. Chem. Soc.* **52**, 5289 (1930).

15. W. H. Carothers and J. W. Hill, *J. Amer. Chem. Soc.* **54**, 1566 (1932).
16. W. H. Carothers, U.S. Patent 2,071,250 (1937).
17. W. H. Carothers, U.S. Patent 2,130,523 (1938).
18. W. H. Carothers, U.S. Patent 2,130,947 (1938).
19. W. H. Carothers, U.S. Patent 2,130,948 (1938).
20. H. K. Livingston, M. S. Sioshansi, and M. D. Glick, *Macromol. Sci.—Revs. Macromol. Chem.* **6** (1), **29** (1971).
21. V. L. Shashoua, W. W. Sweeny, and R. F. Tietz, *J. Am. Chem. Soc.* **82**, 866 (1960).
22. R. J. Gaymans and E. H. J. P. Bour, Netherlands Patent Applic. 80 01,763 and 80 01,764 (10/16/81); E. H. J. P. Bour and J. M. M. Warnier, European Patent Appli. 77, 106 (4/20/83).
23. Japanese Patent Ko Kai-Tokkyo Kho 59 88,910 (5/23/84); Japanese Patent Kohai Tokkyo Kho 59 76,914 (5/2/84).
24. J. Preston, in *Encycl. Polym. Sci. Eng.* **11**, 381 (1988); J. W. Hannell, *Polym. News* **1** (1), **8** (1925); C. W. Stephens, U.S. Patent 3,049,518 (8/14/62); F. W. King, U.S. Patent 3,079,219 (2/26/63).
25. R. E. Wilfong and J. Zimmerman, *J. Applied Polym. Sci.* **17**, 2039 (1973).
26. W. Sweeny and J. Zimmerman, *Encyclo. Polym. Sci. Technol.* **10**, 483 (1961).
27. J. Zimmerman, *Encyclo. Polym. Sci. Eng.* **11**, 315 (1988).
28. J. Preston, *Encyclo. Polym. Sci. Eng.* **11**, 381 (1988).
29. R. J. Welgos, *Encyclo. Polym. Sci. Eng.* **11**, 410 (1988).
30. P. W. Morgan, "Condensation Polymers: By Interfacial and Solution Methods," Wiley (Interscience), New York, 1965.
31. P. W. Morgan and S. L. Kwolek, *J. Polym. Sci.* **2**, 181 (1964).
32. E. L. Wittbecker and P. W. Morgan, *J. Polym. Sci.* **40**, 289 (1959).
33. P. W. Morgan and S. L. Kwolek, *J. Chem. Educ.* **36**, 182 (1959).

5
POLYMERIZATION OF ALDEHYDES

In the polymerization of aldehydes the reaction shown in Eq. (1) occurs to give acetal resins. These polymers may be considered polyethers or polyacetals. This chapter will not consider in detail polyacetals prepared by the reaction of aldehydes with polyols as described in Eq. (2).

$$n\text{RCH}=\text{O} \longrightarrow \left[\begin{array}{c} \text{R} \\ | \\ \text{CH}-\text{O} \end{array}\right]_n \quad (1)$$

$$n\text{RCH}=\text{O} + n\text{HOR'OH} \xrightarrow{n\text{H}_2\text{O}} \left[\begin{array}{c} \text{R} \\ | \\ \text{R'OCH}-\text{O} \end{array}\right]_n \quad (2)$$

The first polymerization of an aldehyde to be reported was that of formaldehyde in 1859 [1]. Butlerov [1] erroneously designated the structure of this material as a dimer, $(\text{CH}_2\text{O})_2$. Its structure was later investigated by Hofmann [2], Tollens and Mayer [3], Lösekann [4], and Delepine [5]. Staudinger [6] confirmed Delepine's suggestion that these materials are high molecular weight substances. Staudinger and co-workers [7] also succeeded in preparing high molecular weight polymers of formaldehyde. They also found that end-capping helped to prevent hemiacetal degradation. These polymers are now known as polyoxymethylenes. In 1942 E. I. Du Pont de Nemours & Co. [9–11] was awarded the first of many patents describing high molecular weight polyoxymethylenes derived by the homopolymerization of formaldehyde, polymers which are now known under the trade name Delrin [11]. Celanese is marketing a

5 POLYMERIZATION OF ALDEHYDES

polyacetal copolymer under the trade name Celcon; it is prepared by the copolymerization of trioxane [12–17] and ethylene oxide.

These acetal resins are in the category of engineering plastics since they have excellent mechanical, chemical, and electrical properties [18]. The acetal resins find use in automotive components, plumbing hardware (sheets, rods, and tubes), electrical components, appliances, consumer products (aerosol containers), and machinery parts (glass bearings). The acetal resins compete for end use application with metals, polyamides (Nylon), polycarbonates, acrylonitrile-butadiene-styrene (ABS), polysulfones, polyphenylene oxides, and polyimides.

The resins are used predominantly for injection molding and extrusion operations. The resins are also available in grades useful for preparing glass or fluorocarbon-reinforced fibers.

United States consumption of acetal resins was estimated to reach 220 million pounds by 1994, up from 10 million pounds in 1961 [19].

Other polyaldehydes have also been prepared from acetaldehyde [20, 21], butyraldehyde [22, 23], chloral [24–27], monochloroacetaldehyde [27, 28], dichloroacetaldehyde [27, 29], and other aldehydes [30, 31].

Chloral has been reported to be polymerized by a monomer casting method using an anionic initiator to give polychloral [24]. The polymer does not support combustion [32] in oxygen and does not melt or drip but decomposes. Polychloral degrades at 200°C and gives mainly the monomer when this degradation is carried out under reduced pressure.

The mechanism of the cation-catalyzed polymerization of acetaldehyde has been suggested to involve generation of a carbonium ion and addition to the carbonyl oxygen atom of each aldehyde group [33].

$$CH_3CH{=}O + BF_3 \longrightarrow CH_3\overset{H}{\underset{}{C}}{=}O{:}BF_3 \longrightarrow \left[{}^+\overset{CH_3}{\underset{H}{C}} - \bar{O}BF_3 \right] \xrightarrow{CH_3CH{=}O}$$

$$\left[BF_3^- O - \overset{CH_3}{\underset{H}{C}} - O - \overset{CH_3}{\underset{H}{C}}{}^+ \right] \xrightarrow{nCH_3CH{=}O} BF_3^- O - \overset{CH_3}{\underset{H}{C}} {\left[-O - \overset{CH_3}{\underset{H}{C}} - \right]}_n O - \overset{CH_3}{\underset{H}{C}}{}^+$$

(3)

Polyacetaldehyde degrades readily to acetaldehyde unless the polymer is treated with acetic anhydride and pyridine [30]. Etherification with ortho esters is also effective. The addition of antioxidants and the addition of thermal stabilizers of the amide type give a further improvement in stability [30]. These polymers to date have not found any commercial use because they still do not have enough thermal stability.

Acetaldehyde polymers *without* a polyether structure have been reported by TsuiTsui and co-workers using a tetraisopropyl titanate catalyst [34].

5-1. Preparation of Polyacetaldehyde by Cationic Polymerization of Acetaldehyde Using BF$_3$ Catalysis [35]

$$n\text{CH}_3\text{CHO} \xrightarrow[\substack{\text{ethylene} \\ -130°C}]{\text{BF}_3 \text{ etherate}} -\text{O}-\underset{\text{H}}{\overset{\text{CH}_3}{\text{C}}}-\text{O}-\underset{\text{H}}{\overset{\text{CH}_3}{\text{C}}}-\text{O}-\underset{\text{H}}{\overset{\text{CH}_3}{\text{C}}}-\text{O}-\underset{\text{H}}{\overset{\text{CH}_3}{\text{C}}}-\text{O}-\underset{\text{H}}{\overset{\text{CH}_3}{\text{C}}}-\text{O}-\underset{\text{H}}{\overset{\text{CH}_3}{\text{C}}}- \quad (4)$$

■ **CAUTION:** Use a well-ventilated hood because aldehydes are toxic.

Into a 1 liter round-bottomed flask equipped with a mechanical stirrer, nitrogen inlet and outlet, and a $-200°C$ thermometer is condensed by means of liquid nitrogen 250 ml of ethylene (b.p. $-104°C$). Then 48 gm (1.2 moles) of acetaldehyde (see Note) is slowly injected through a serum cap using a hypodermic syringe, and stirring is begun. The temperature is kept at $-130°C$ and then 5 drops of BF$_3$ etherate is injected. After 15–20 min the reaction mixture increases in viscosity. It is allowed to react for 1 hr. Next 100 ml of pyridine is slowly added at $-104°C$ to deactivate the catalyst and dissolve the polymer. The pyridine solidifies and the reaction is kept at $-78°C$ overnight in the hood to allow the ethylene to evaporate slowly. Then a solution of 20 ml of pyridine and 200 ml of acetic anhydride is added to the pyridine solution of the polymer and the mixture stirred for 2 hr. The polymer is isolated by precipitating it in ice–water. The polymer is kneaded by hand in water using rubber gloves, to destroy all the excess acetic anhydride. The polymer is dissolved in ether, extracted several times with water, dried, filtered, and concentrated under reduced pressure to afford 44 gm (92%) of polyacetaldehyde, $\eta_{inh} = 3.67$ (in 0.1% butanone solution at 25°C).

NOTE: The acetaldehyde is twice distilled in a low-temperature still over an antioxidant (0.1% β,β'-dinaphthyl-p-phenylenediamine).

Vogl [36-39] reported that acetaldehyde polymerizes to give a stereoregular crystalline polymer (isotactic) using anionic catalysts such as alkali alkoxides (lithium sec-butoxide or potassium triphenylmethoxide), LiAlH$_4$, and alkali metal alkyls (n-BuLi). The polymerization is carried out in such solvents as toluene, propylene, and isobutylene at $-40°$ to $-80°C$.

5-2. Polymerization of Acetaldehyde Using Potassium Triphenylmethoxide Catalysis [39]

$$n\text{CH}_3\text{CH}=\text{O} \xrightarrow{(C_6H_5)_3\text{COK}} \left[\begin{array}{c} \text{CH}_3 \\ | \\ \text{CH}-\text{O} \end{array}\right]_n \qquad (5)$$

A 500 ml three-necked, round-bottomed flask is equipped with a thermometer, stirrer, nitrogen inlet and outlet, and serum stopper. Nitrogen gas is added to flush the flask out and then reduced to such a rate as to keep a static nitrogen atmosphere. The flask is immersed in a dry ice–acetone bath and 200 ml of propylene is condensed in. When the temperature in the flask reaches $-75°C$, 5 ml of 0.125 N potassium triphenylmethoxide (see Note) in benzene*–heptane is added using a hyperdermic syringe. While the reaction mixture is being vigorously stirred, 44 gm (1.0 mole) of acetaldehyde is added with a hyperdermic syringe. The polymer starts to precipitate as a gelatinous polymer and after 1 hr 200 ml of cold acetone containing 0.5 ml of acetic acid is added. The polymer is filtered after the reaction mixture warms to room temperature and 31.7 gm (72%) is obtained. Approximately 45% of this material is crystalline. The polymer may be stabilized by end-capping with pyridine and acetic anhydride as described previously.

NOTE: Potassium triphenylmethoxide is prepared as follows: To a 500 ml dry three-necked flask equipped with a dropping funnel, condenser, and drying tube, and magnetic stirrer is added 2.2 gm of potassium sand in 10 ml of heptane. Then 50 ml of dry benzene* is added followed by the dropwise addition (2 hr) of 13 gm (0.050 mole) of triphenylmethanol dissolved in 250 ml of benzene*. Next the reaction mixture is boiled for 0.5 hr while the original blue color changes to tan. The mixture is filtered under nitrogen into a dry flask and is ready for use. (*CAUTION: Benzene is carcinogenic.)

Vogl [38, 40], Natta [41], and Furukawa [42, 43] and co-workers reported independently that crystalline polyacetaldehyde is obtained by polymerizing acetaldehyde at low temperature in the presence of organometallic compounds or metal alkoxides. The infrared spectrum of the crystalline polymer gives sharper absorption bands with more distinct relative intensities. The X-ray diffraction diagrams indicate that the polymer is definitely crystalline.

REFERENCES

1. A. M. Butlerov, *Ann. Chem. Pharm.* **111**, 242 (1859).
2. A. W. Hofmann, *Ann. Chem. Pharm.* **145**, 357 (1868); *Ber. Deut. Chem. Ges.* **2**, 156 (1869).
3. B. Tollens and F. Mayer, *Ber. Deut. Chem. Ges.* **21**, 1566, 2026, and 3503 (1888).
4. G. Lösekann, *Chem.-Ztg.* **14**, 1408 (1890).
5. M. Delepine, *C. R. Acad. Sci.* **124**, 1528 (1897); *Ann. Chim. Phys.* [4] **15**, 530 (1898).
6. H. Staudinger, "Die Hochmolekularen Organischen Verbindungen," pp. 224–287, Springer-Verlag, Berlin and New York, 1960 (new ed.).
7. H. Staudinger and A. Gaule, *Ber. Deut. Chem. Ges.* **49**, 1897 (1916); H. Staudinger and R. Signer, *Helv. Chim. Acta* **11**, 1847 (1958); H. Staudinger, R. Signer, H. Johner, O. Schweitzer, M. Lüthy, W. Kern, and D. Fussidis, *Justus Liebigs Ann. Chem.* **474**, 145 (1929).
8. P. R. Austin and C. E. Frank, U.S. Patent 2,296,249 (1942); R. N. MacDonald, U.S. Patent 2,828,286 (1958); H. H. Goodman, Jr., U.S. Patent 2,994,687 (1961).
9. R. N. MacDonald, U.S. Patent 2,768,994 (1956).
10. C. E. Schweitzer, R. N. MacDonald, and J. O. Punderson, *J. Appl. Polym. Sci.* **1**, 158 (1959).
11. J. C. Bevington and H. May, *Encycl. Polym. Sci. Technol.* **1**, 609 (1964).
12. A. K. Schneider, U.S. Patent 2,795,571 (1957).
13. K. W. Bartz, U.S. Patent 2,947,728 (1960).
14. O. H. Axtell and C. M. Clarke, U.S. Patent 2,951,059 (1960); C. J. Bruni, U.S. Patent 2,989,510 (1961); C. L. Michaud, U.S. Patent 2,982,758 (1961); Farbwerke Hoechst, A. G., Australian Patent Appl. 55,706/59 (1960); British Industries Plastics Ltd., Belgian Patent 592,599 (1961); D. E. Hudgin and F. M. Berardinelli, French Patent 1,216,327 (1960); G. J. Bruni, French Patent 1,226,988 (1960).
15. D. E. Hudgin and F. M. Berardinelli, U.S. Patents 2,989,505–2,989,509 (1961).
16. S. O. Zismann, G. I. Fáidel, and P. G. Konovalov, Russian Patent S U 221284 (1979); *Chem. Abstr.* **91**, 5710v (1979).

17. I. Glavchev, V. Kabaivanov, and M. Natov, *God. Vissh. Khimikotekhnol. Inst. Sofia* (Bulgaria) **17** (1), 207 (1973), *Chem. Abstr.* **82**, 98537w (1973).
18. D. Oosterhof, "Chemical Economics Handbook," Acetal Resins, No. 580,0121A and 580,0122A, Dec. 1971.
19. D. Oosterhof, "Chemical Economics Handbook," Acetal Resins No. 580,0121C, Dec. 1971.
20. M. S. Travers, *Trans. Faraday Soc.* **32**, 246 (1936).
21. M. Letort, *C. R. Acad. Sci.* **202**, 767 (1936).
22. J. B. Conant and W. R. Peterson, *J. Amer. Chem. Soc.* **52**, 1659 (1930).
23. A. Novak and E. Whalley, *Can. J. Chem.* **37**, 1710 and 1718 (1959).
24. Chemical & Engineering News, *Chem. Eng. News* **52** (12), 41 (1972).
25. J. Böeseken and A. Schimmel, *Rec. Trav. Chim. Pays-Bas* **32**, 112 (1913).
26. S. Gaertner, *C. R. Acad. Sci.* **1**, 513 (1906).
27. I. Rosen, *Polym. Prepr., Amer. Chem. Soc., Div. Polym. Chem.* **7**, 221 (1966).
28. K. Natterer, *Monatsh, Chem.* **3**, 442 (1882).
29. O. Jacobsen, *Ber. Deut. Chem. Ges.* **8**, 87 (1875).
30. O. Vogl, *Polym. Prepr., Amer. Chem. Soc., Div. Polym. Chem.* **7**, 216 (1966).
31. J. Furukawa and T. Saegusa, "Polymerization of Aldehydes and Oxides," Wiley (Interscience), New York, 1963.
32. V. V. Korshak and A. L. Rusanov, *Usp. Khim.* **58** (6), 1006 (1989); see also *Chem. Abstr.* **111**, 214955z (1989).
33. O. Vogl and W. M. D. Bryant, *J. Polym. Sci., Part A* **2**, 4633 (1964).
34. M. Tsutsui, M. Makita, and B. Gorewit, *J. Polym. Sci., Polym. Chem. Ed.* **14** (5), 35 (1976).
35. O. Vogl, *J. Polym. Sci., Part A* **2**, 4591 (1964).
36. O. Vogl, *J. Polym. Sci.* **46**, 261 (1969).
37. O. Vogl, *J. Polym. Sci., Part A* **2**, 4607 (1964).
38. O. Vogl, *J. Polym. Sci., Part A* **2**, 4621 (1964).
39. O. Vogl, German Patent 1,144,921 (1963).
40. O. Vogl, Belgian Patent 580,553 (1959).
41. G. Natta, G. Mazzanti, P. Coeradini, and I. W. Bassi, *Makromol. Chem.* **37**, 156 (1960).
42. J. Furukawa, T. Saegusa, H. Fujii, A. Kawasaki, H. Imai, and Y. Fujii, *Makromol. Chem.* **37**, 149 (1960).
43. J. Furukawa, T. Saegusa, and H. Fujii, *Makromol. Chem.* **44–46**, 398 (1960).

6
POLYMERIZATION OF EPOXIDES AND CYCLIC ETHERS

Ethylene oxide, various epoxides, and cyclic ethers can be polymerized with anionic, cationic, and coordination type catalysts.

Wurtz [1] was the first to report the base-initiated polymerization of ethylene oxide. Roithner [2] extended this work and showed that raising the temperature accelerated the polymerization (degree of polymerization approx. 30). In addition he showed that aqueous alkali was also an effective catalyst. Wurtz [3] also showed that polyoxyethylene acetates could be prepared by heating ethylene oxide with acetic anhydride. Staudinger [4, 5] later reinvestigated the polymerization of ethylene oxide and systematically studied the effect of various anionic and cationic catalysts on the polymerization. For example, the most effective catalysts found were $(CH_3)_3N$, Na, K, and $SnCl_4$. Other compounds showing catalytic activity were $NaNH_2$, ZnO, SrO, and CaO. The latter catalysts gave a degree of polymerization varying from 250 to 2500. Activated alumina was also effective but the following were ineffective: ultraviolet rays, Florida earth, ferric oxide, magnesium oxide, lead oxide, silica gel, and activated carbon.

Staudinger [5] also demonstrated that as the molecular weight of polyoxyethylene increased (to 100,000) the melting points (40°–175°C) and the viscosity of a 1% aqueous solution of the polymer increased but the polymer solubility in water tended to diminish (especially at molecular weight 100,000).

In 1957 Hill [6] and co-workers of Union Carbide [7] reported on the commercial process of polymerizing ethylene oxide with specially prepared calcium carbonates and calcium amides.

6 POLYMERIZATION OF EPOXIDES AND CYCLIC ETHERS 47

Hill, Bailey, and Fitzpatrick [6] found that strontium carbonate requires a certain amount of water to be active as a catalyst for the polymerization of ethylene oxide. Samples of strontium carbonates with the required water content were inactive catalysts if they contained nitrate, chlorate, thiosulfate, or tetraborate ions [6].

Pruitt and Baggett [8] of the Dow Chemical Company reported that the reaction product of ferric chloride and propylene oxide afforded a coordination catalyst capable of polymerizing propylene oxide to a high molecular weight crystalline or amorphous rubbery polymer.

The polymerization of propylene oxide to a high molecular weight product has also been reported to occur with a variety of metal alkoxides, halides, and alkyls [9–11].

Low molecular weight polyoxypropylene glycols can be obtained by the potassium hydroxide-catalyzed polymerization of propylene oxide [12]. The latter products are suitable intermediates for the preparation of flexible polyurethane foams, elastomers, and adhesives.

The mechanisms of polymerization for epoxides and cyclic ethers may be summarized as in Scheme 1, using ethylene oxide as an example. Routes (A) through (C) illustrate the following mechanisms, respectively: cationic, anionic, and coordinated anionic–cationic. The coordinate mechanism probably involves two metal atoms and will be discussed later in greater detail.

Today polyethylene glycols are available in a range from water-soluble viscous liquids (mol. wt. below 700) to grease and waxes (mol. wt. 1000 to 20,000) from a variety of sources [13, 14]. Products of molecular weight of about 100,000 to 5,000,000 are available from Union Carbide under the Polyox trademark [14]. The latter have melting points of about $65° \pm 2°C$ and a specific gravity of 1.21 gm/cm^3.

Very low molecular weight polyoxyethylene glycols, $HO(CH_2—CH_2O)_nH$ where $n = 2$–6 are useful starting materials for the preparation of diesters and polyesters.

6-1. Preparation of Polyoxypropylene Glycol by the Room-Temperature Polymerization of Propylene Oxide Using KOH [15]

To a flask equipped with a mechanical stirrer and condenser is added 200 gm (3.45 moles) of dried propylene oxide (dried over

SCHEME 1
Outline of the Mechanisms of Polymerization of Alkylene Oxides

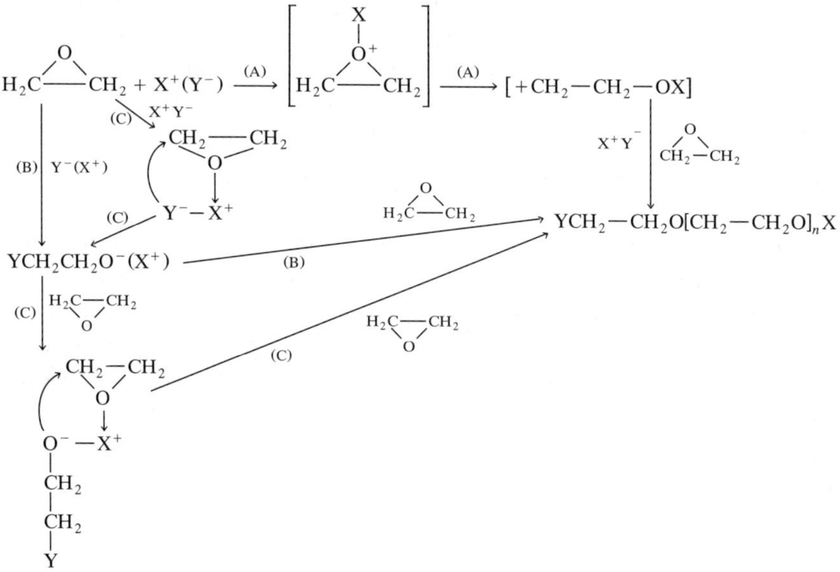

$$H_2C\overset{O}{\underset{}{\diagup\!\!\!\diagdown}}CH_2 + X^+(Y^-) \xrightarrow{(A)} \left[H_2C\overset{\overset{X}{\underset{}{|}}\overset{+}{O}}{\underset{}{\diagup\!\!\!\diagdown}}CH_2\right] \xrightarrow{(A)} [+CH_2-CH_2-OX]$$

(with branches labeled (B), (C) leading to $YCH_2-CH_2O[CH_2-CH_2O]_nX$)

TABLE I
Room Temperature Polymerization of Propylene Oxide with Powdered Potassium Hydroxide[a]

Monomer (gm)	KOH (gm)	Time (hr)	Yield (gm)	OH^- (Eq./gm $\times 10^4$)	$C=C$ (Eq./gm $\times 10^4$)	η_{sp}/c^b	$n^{20}D$
200	40	42	160[c]	3.82	—	—	—
150	20	64	36.5[d]	2.96	—	0.141	—
150	10	48	—	4.88	—	0.138	—
420	10	141	250[d]	—	1.35	0.154[e]	1.4509
406	30	42	209[d]	3.61	—	0.136	1.4509
800	60	220	360[d]	6.02	1.37	0.161	—

[a] Reprinted from St. Pierre and Price [11]. Copyright 1956 by the American Chemical Society. Reprinted by permission of the copyright owner.
[b] 4% solution in benzene; c = gm of polymer in 100 gm of solution.
[c] Isolated by washing product in ether with hydrochloric acid and evaporating to 150°C at 3 mm.
[d] Isolated by shaking an ether solution with Amberlite IR-120(H) to remove base, followed by vacuum evaporation.
[e] Cryoscopic molecular weight (benzene) 3440.

KOH and distilled before use) and 40 gm (1.4 mole) of finely powdered technical grade potassium hydroxide. The mixture is stirred for 42 hr at room temperature and then ether is added to dissolve the polymer. Hydrogen chloride is then bubbled into the solution and the precipitated KCl filtered. The base (KOH) can also be removed by shaking the ether solution with Amberlite IR-120(H) (Rohm & Haas Co.). The filtrate is concentrated under reduced pressure (150°C at 3 mm Hg) to afford 160 gm (80%) of product. This and several other related experiments are summarized in Table I.

6-2. Polymerization of *l*-Propylene Oxide with Solid KOH Catalyst [16]

$$\overline{OH} + CH_2 \underset{O}{-} CH_* \xrightarrow{CH_3} HOCH_2 \underset{*}{CH} - O^- \longrightarrow$$

$$HOCH_2 \underset{*}{\overset{CH_3}{CH}} - OCH_2 - \underset{*}{\overset{CH_3}{CHO^-}} \longrightarrow HO \left[CH_2 \underset{*}{\overset{CH_3}{CH}} - O \right]_n H \quad (1)$$

To a cooled Pyrex tube is added 2.5 gm (0.043 mole) of *l*-propylene oxide and 0.5 gm (0.00895 mole) of powdered potassium hydroxide. The tube is flushed with nitrogen, sealed, and put in a shaker water bath at 25°C for 50 hr. The mixture finally solidifies to a wax after 50 hr and then is dissolved in 250 ml of benzene. The benzene solution is washed successively with water, dilute aqueous sulfuric acid, aqueous sodium bicarbonate solution, and then with distilled water until neutral. The benzene solution is freeze-dried (under reduced pressure) to afford 2.2 gm (88%) of a white crystalline poly(*l*-propylene oxide), m.p. 55.5°–56.5°C. (**CAUTION:** Benzene is carcinogenic.)

Using similar conditions *dl*-propylene oxide affords 88% of a liquid poly(*dl*-propylene oxide).

6-3. Polymerization of Ethylene Oxide Using SnCl$_4$ (or BF$_3$) Catalyst [17]

*Materials.** Ethylene oxide is purified by low-temperature distillation, by distillation at $-78°C$ from dried sodium hydroxide pellets, and finally by drying over barium oxide, all under vacuum.

Stannic chloride is fractionally distilled from tin metal, then twice distilled under vacuum.

Solvents are fractionated from phosphorus pentoxide. Ethylene chloride is treated with aluminum chloride, washed and dried before fractionating.

Method.* Polymerizations are carried out in an all-glass vacuum system and the rate of reaction in ethylene chloride solution is followed by the decrease in vapor pressure of the reaction mixture. The relationship between oxide concentration and vapor pressure is linear and hence readily determined from known mixtures.

Stannic chloride and water are measured out by allowing the liquid, at controlled temperature, to evaporate into bulbs of known volume until equilibrium is established. The quantity is then determined from the vapor pressure of the liquid.

Ethylene oxide and ethyl chloride are measured in the gas phase, but the less volatile solvents are pipetted into a trap containing phosphorus pentoxide and degassed before distilling into the reaction vessel.

Catalyst, water if any, and solvent in that order are condensed into the reaction vessel and brought to reaction temperature. Ethylene oxide is then rapidly distilled in from a small side arm on the main reaction vessel. Magnetic stirring establishes equilibrium within 3 minutes.

Ethyl chloride is employed as solvent when reaction products are to be determined because it can be separated readily by distillation. Ethylene chloride is used for most of the kinetic studies because of its lower vapor pressure.

Reaction products are isolated by treating a 10% solution of ethylene oxide with 1 mole % stannic chloride at 20°C. Solvent, excess oxide and volatile products are removed by distillation, leaving a white wax having an infrared spectrum identical with that of a polyethylene glycol of similar melting point. The volatile product is found, by mass spectrometer analysis, to consist of 92% dioxane and 8% 2-methyl-1,3-dioxolane.

The same technique is also used for the polymerizations with BF_3 [17]. Decreasing the temperature in the latter case raised the molecular weight from 760 at 20°C to 880 at 0°C and 925 at −20°C.

---

*From Worsfold and Eastham [17]. Copyright 1957 by the American Chemical Society. Reprinted by permission of the copyright owner.

Molecular weights.[†] Molecular weights are determined viscometrically in ethylene chloride, using Ostwald viscometers with flow times of about 200 sec at 20°C. For molecular weights less than 1000, the relationship

$$\eta_{sp}/c = 0.048 + 0.000204\,M$$

is used where c is in moles per liter. This expression is obtained using commercial polyglycols whose molecular weights are determined by end-group analysis and by the depression of the benzene freezing point.

Molecular weights above 1000 are given by the expression

$$\eta_{sp}/c = 0.160 + 0.000089\,M$$

which is obtained from commercial polyglycols whose molecular weights are first determined in carbon tetrachloride using the viscometric data of Fordyce and Hibbert [18].

Polymer samples are obtained from the reaction mixture by quenching aliquots in aqueous alcohol, evaporating on a steam bath, then drying for 2 or 3 hr under vacuum at 70°C.

REFERENCES

1. A. Wurtz, *C. R. Acad. Sci.* **49**, 813 (1859); *Chem. Ber.* **10**, 90 (1877); *Bull. Soc. Chim. Fr.* [2] **29**, 530 (1878).
2. E. Roithner, *Monatsh. Chem.* **15**, 679 (1894); *J. Chem. Soc., London* **68**, 319 (1895).
3. A. Wurtz, *Ann. Chim. Phys.* [3] **69**, 334 (1863).
4. H. Staudinger and O. Schweitzer, *Chem. Ber.* **63**, 2395 (1929).
5. H. Staudinger and H. Lehmann, *Justus Liebigs Ann. Chem.* **505**, 41 (1933).
6. F. N. Hill, F. E. Bailey, Jr., and J. T. Fitzpatrick, *Ind. Eng. Chem.* **50**, 5 (1958).
7. Union Carbide and Carbon Corp., Australian Patent Appl. 27792 (1957).
8. M. E. Pruitt and J. M. Baggett, U.S. Patent 2,706,181 (1955).
9. A. B. Borkovec, U.S. Patent 2,873,258 (1959).

[†]From Worsfold and Eastham [17]. Copyright 1957 by the American Chemical Society. Reprinted by permission of the copyright owner.

10. M. Osgan and C. C. Price, *J. Polym. Sci.* **34**, 153 (1959); R. O. Colclough, G. Geoffrey, and A. H. Jagger, *Ibid.* **48**, 273 (1960).
11. P. E. Ebert and C. C. Price, *J. Polym. Sci.* **34**, 157 (1959).
12. D. M. Simons and J. J. Vebanc, *J. Polym. Sci.* **44**, 303 (1960).
13. "Carbowax Polyethylene Glycol," Technical Bulletin F4772F, Union Carbide Corporation, Chemical and Plastics Division, New York, 1970; "Dow Polyethylene Glycols," Technical Bulletin 125-230-59, Dow Chemical Company, Midland, Michigan, 1959; "Polyethylene Glycols," Technical Bulletin, Jefferson Chemical Company, Houston, Texas, 1961; "Polyethylene Glycols," Technical Bulletin OC-107-1061, OC-108-1061, Olin Mathieson Chemical Corporation, Organic Chemicals Division, 1961; Pluracol E Technical Data, Wyandotte Chemical Corporation, Wyandotte, Michigan, 1960.
14. "Polyox Water Soluble Resins," Technical Bulletin F-40246E, Union Carbide Corporation, Chemical and Plastics Division, New York, 1968.
15. L. E. St. Pierre and C. C. Price, *J. Amer. Chem. Soc.* **78**, 3432 (1956).
16. C. C. Price and M. Osgan, *J. Amer. Chem. Soc.* **78**, 4787 (1956).
17. D. J. Worsfold and A. M. Eastham, *J. Amer. Chem. Soc.* **79**, 897 (1957).
18. R. Fordyce and H. Hibbert, *J. Am. Chem. Soc.* **61**, 1912 (1939).

7
POLYUREAS

Polyureas are obtained by the reactions involving di- and polyfunctional starting materials, as shown in Eq. (1). The reaction of diisocyanates with water can also be used since the diamine can be formed *in situ* and reacted with the diisocyanates to give the polyurea. This reaction is employed in the manufacture of foams, and catalysts are usually employed [1, 2].

Polyureas, like polyamides, have many sites available for hydrogen bonding. The polyureas generally are higher melting and less soluble than the related polyamides. Polyureas are rarely made by melt polymerization techniques because they are thermally unstable at temperatures above 200°C and must be made by solution polymerization techniques.

$$\left[R'-NHCNH \right]_n \underset{X=Cl_2 \text{ or } S}{\overset{COX}{\longleftarrow}} H_2NR'NH_2 \overset{NH_2-C-NH_2}{\longrightarrow} \left[R'-NHCNH \right]_n$$

$$\downarrow R(NCX)_2, X=O, S \text{ or } R(NHCOOC_2H_5)_2$$

$$\left[R'-NHCNHRNHC-NH \right]_n \quad (1)$$

- **CAUTION:** All reactions should be carried out in a well-ventilated hood with the use of proper personal protection. It is recommended that all Material Safety Data Sheets (MSDS) be obtained for each chemical being used and read very carefully by all those engaged with the synthesis procedures.

1. REACTION OF DIAMINES WITH DIISOCYANATES

Diamines are frequently used to cure diisocyanates, especially isocyanate prepolymers, to give polyureas. The reaction has the advantages that no by-products are produced and that it can be carried out at low temperatures. This process is related to the diamine–phosgene reaction in that here the diisocyanate must first be isolated and reacted further with diamines. Co-polymers are easily formed in this latter type of reaction when the reaction components have different carbon backbones.

Aromatic diisocyanates are much more reactive (reaction time: approx. 3–10 min) [3, 4] than aliphatic diisocyanates (reaction time: approx. 1–5 hr) [5–9] with amines to give polyureas. In addition aromatic diamines are more reactive than aliphatic diamines with diisocyanates.

The use of tertiary amines [10], organotin compounds [11], stannic chloride [3], and lithium chloride [3] has been found to improve the reactivity of N-alkylamines, which otherwise react very sluggishly with diisocyanates.

7-1. Preparation of a Polyurea from Bis(3-aminopropyl) Ether and 1,6-Hexamethylene Diisocyanate [6]

■ **CAUTION:** Isocyanates and diisocyanates are very toxic and must be used with extreme care in a well-ventilated hood. Please read all Material Safety Data Sheets (MSDS) and follow recommended precautions. The amines also are toxic and must be used with similar safety precautions.

$$(CH_2)_6(NCO)_2 + \bar{C}lH_3\overset{+}{N}-(CH_2)_3-O-(CH_2)_3-\overset{+}{N}H_3Cl^- \xrightarrow[H_2O]{NaOH}$$

$$\left[-(CH_2)_6-NH\overset{O}{\overset{\|}{C}}NH(CH_2)_3-O-(CH_2)_3NH\overset{O}{\overset{\|}{C}}NH-\right]_n + 2NaCl \quad (2)$$

To a flask containing a mixture of 168 gm (1.0 mole) of 1,6-hexamethylene diisocyanate is added 132 gm (1.0 mole) of bis(3-

aminopropyl) ether dissolved in a mixture of 1.3 liters of water and 2 liters of 1 N HCl. Then 18 gm of benzyl p-hydroxybiphenyl polyglycol is added as an emulsifier. Next 2 liters of ice cold 1 N NaOH is added over a 30 min period with stirring and cooling to form the polyurea. The polyurea is filtered to afford 234 gm (78%), m.p. 225°–227°C. At 26°C the polyurea could be drawn into threads which could be stretched in the cold and easily dyed.

2. REACTION OF DIAMINES WITH UREA

Urea reacts with diamines at 130°C to give ammonia and polyureas (Eq. 3). The polycondensation can be carried out in the melt or in a solvent such as m-cresol [12, 13] or n-alkylpyrrolidones [14]. In some cases tertiary amine catalysts in dioxane solvent are used to accelerate the polymerization [15].

$$NH_2RNH_2 + NH_2CONH_2 \xrightarrow{-2NH_3} +NHCONHR+_n \qquad (3)$$

A variety of aliphatic and aromatic primary diamines have been used for the polymerization in addition to N-substituted diamines [16], unsaturated diamines [17], and heterocyclic-based diamines [18, 19].

The preparation of several polymethylene straight-chain polyurea homopolymers have been reported, for example, $+CONH-(CH_2)_n-NH+_x$ where $n = 3$ [20], 6 [21, 22], 7 [23–27], 10 [21, 25, 26], and 12 [26].

7-2. Preparation of Poly(4-oxyhexamethyleneurea) [17]

$$NH_2\overset{O}{\overset{\|}{C}}NH_2 + H_2N(CH_2)_3-O-(CH_2)_3NH_2 \xrightarrow[\text{heat}]{N_2}$$

$$\left[+(CH_2)_3-O-(CH_2)_3-NH-\overset{O}{\overset{\|}{C}}-NH+_n\right] + 2NH_3 \qquad (4)$$

To a test tube with a side arm are added 7.5 gm (0.125 mole) of urea and 16.5 gm (0.125 mole) of bis(γ-aminopropyl) ether. A capillary tube attached to a nitrogen gas source is placed on the bottom of the tube and the temperature is raised to 156°C and kept there for 1 hr, during which time ammonia is evolved. The temperature is raised to 231°C for 1 hr and then to 255°C for an additional hour. At the end of this time a vacuum is slowly applied to remove the last traces of ammonia.

■ CAUTION: Frothing may be a serious problem if the vacuum is applied too rapidly.) The polymer is cooled, the test tube broken, and the polymer isolated. The polymer melt temperature is approximately 190°C, and the inherent viscosity is approximately 0.6 in m-cresol (0.5% conc., 25°C).

REFERENCES

1. S. R. Sandler and W. Karo, "Organic Functional Group Preparations," Vol. 2, 2nd ed., Chapter 6, Academic Press, New York, 1986.
2. H. G. J. Overmars, *Encycl. Polym. Sci. Technol.* **11**, 464 (1969); O. E. Snider and R. J. Richardson, *ibid.*, p. 495; M. A. Dietrich and H. W. Jacobson, U.S. Patent 2,709,694 (1955).
3. M. Katz, U.S. Patent 2,975,157 (1961).
4. M. Katz, U.S. Patent 2,888,438 (1959).
5. H. V. Boening, N. Walker, and E. H. Meyers, *J. Appl. Polym. Sci.* **5**, 384 (1961).
6. W. Lehmann and H. Rinke, U.S. Patent 2,852,494 (1958).
7. I. Yasuzawa, M. Yamaguchi, and Y. Minoura, *J. Polym. Sci., Polym. Chem. Ed.* **17**, 3387 (1979).
8. P. Schlack, German Patent 920,511 (1954).
9. W. Lehmann and H. Rinke, U.S. Patent 2,761,852 (1956).
10. O. Y. Fedotova, A. G. Grozdow, and I. A. Rusinovskaya, *Vysokomol. Soedin.* **7**, 2028 (1965).
11. S. R. Sandler and F. R. Berg, *J. Appl. Polym. Sci.* **9**, 3909 (1965).
12. W. R. Grace Co., British Patent 863,29 (1961).
13. Y. Inaba and K. Ueno, Japanese Patent 10,092 (1956).
14. R. Gabler and H. Müller, U.S. Patent 3,185,656 (1965).
15. Y. Furuya and K. Itoho, *Chem. Ind. (London)* p. 359 (1967).
16. H. Iiyama, M. Asakura, and K. Kimoto, *Kogyo Kagaku Zasshi* **68**, 236 (1966).
17. G. Kimura, S. Kaichi, and S. Fujigake, *Yuki Gosei Kagaku Kyokai Shi* **23**, 241 (1965).
18. Y. Inaba, K. Miyake, and G. Kimura, U.S. Patent 3,054,777 (1962).
19. F. Veatch and J. D. Idol, Belgian Patent 614,386 (1962).
20. Y. Iwakura, *Chem. High Polym.* **4**, 97 (1947).
21. E. I. Du Pont de Nemours and Co., Inc., British Patent 530,267 (1940).
22. Y. Inaba, K. Miyake, and G. Kimura, U.S. Patent 3,046,254 (1962).

23. R. Gabler and H. Müller, U.S. Patent 3,185,656 (1965).
24. G. Kimura, S. Kaichi, and S. Fujigake, *Yuki Gosei Kagaku Kyokai Shi* **23**, 241 (1965).
25. Y. Inaba, K. Miyake, K. Kimoto, and K. Kimura, French Patent 1,207,356 (1958).
26. P. Borner, W. Gugel, and R. Pasedag, *Makromol. Chem.* **101**, 1 (1967).
27. K. Ueda, H. Mikami, and T. Okawara, Japanese Patent 13,243 (1960).

8
POLYURETHANES

The reaction of isocyanates with alcohols to give *N*-carbamates has been described in detail in Chapter 10 of Volume II of *Organic Functional Group Preparations* [1]. Although polyurethanes were briefly described in that chapter, the subject will be covered here in greater detail, especially from the preparative viewpoint.

Polyurethanes were first reported by Bayer in 1937 [2, 3] and Rinke and co-workers in 1939 [4], to result from the reaction of diisocyanates with dihydric alcohols. From 1945 on, many other patents were issued on the preparation of polyurethanes from the reaction of diamines and bischloroformates at low temperatures [5, 6]. Soon after, several reports appeared describing the high-temperature solution and melt polymerization methods which involve the reaction of diisocyanates with diols [3, 7]. Another method has been reported in which polyurethanes are prepared by the direct reaction of 1,4-dichloro-2-butene with sodium cyanate and diols [8]. However, this latter method may not be of general utility since further research is required. It is presented at this time for information only. These methods are briefly summarized in Scheme 1. Scheme 1 also indicates that ester exchange reactions can also be used to give polyurethanes. This latter reaction is particularly important in one-package type adhesives and coatings.

For more recent developments in the polyurethanes area, one should consult the procedures of the September 30–October 3, 1990, SPI conference in Orlando, Florida [9]. As far as synthesis is

SCHEME 1
Preparation of Polyurethanes

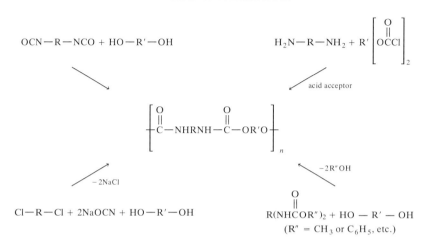

concerned, the following developments should be noted:

1. Replacements for chlorofluorocarbons for polyurethane foam
2. Reaction injection molding (RIM)
3. Polyurethane interpenetrating networks (IPNs)
4. Flame retardants for polyurethanes

- **CAUTION:** Handling and use of isocyanates should be undertaken with great care in order to avoid any exposure. Isocyanates are suspect carcinogens and cause irritation to the respiratory tract (nose, throat, and lungs). The American Conference of Governmental Industrial Hygienists (ACGIH) has set threshold limit values (TLV) for most diisocyanates (0.005 ppm time-weighted coverage over 8 hr [10]). The OSHA permissible exposure limits for TDI and MDI are 0.02 ppm as a ceiling value [11, 12].

In addition, care should be exercised in the handling of the amine or tin catalysts, polyols, and blowing agents [11]. A thorough reading of each Material Safety Data Sheet (MSDS) from the manufacturer or supplier should be standard practice before using these materials.

The combustion of polyurethanes also poses safety hazards since HCN and other toxic combustion products have been reported to evolve therefrom [13].

Polyurethane resins are usually prepared by the reaction of a long-chain diol with an excess of the diisocyanate to obtain a "prepolymer" with terminal isocyanate groups [14] (Eq. 1).

$$R(NCO)_2 + HO-\sim\sim-OH \longrightarrow OCN-R-NHCOO\sim\sim OCONH-R-NCO \quad (1)$$

This "prepolymer" can react separately with diols or diamines of low molecular weight to cause further chain extension or polymerization (curing of the prepolymer). On reaction with water, the "prepolymer" can also be used to give foams, in which amines or tin compounds are used as catalysts in the foaming process. The properties of the polyurethane resin are controlled by the choice of the diisocyanate and polyol. The stiffness of the aromatic portion of the diisocyanate may be increased by using 1,5-naphthalene diisocyanate in place of 2,4-toluene diisocyanate. In addition, the flexibility may be increased by using a polyether polyol derived from propylene oxide or tetrahydrofuran. Using aromatic polyols and triols would lead to chain stiffness and crosslinking in the case of

TABLE I
TYPICAL FLEXIBLE AND RIGID POLYURETHANE FOAM FORMULATIONS[a]

Component	Parts by wt.	
	Flexible (low density)	Rigid
Polyol	100 Polyether triol (MW 3000)	29.0 Polyether polyol
Isocyanate[b]	47.8 TDI 80 : 20	51.0 MDI (polymeric)
Water	3.7	0.5
Surfactant	2.0	0.5
Amine catalyst	0.66	0.6
Stannous octoate	0.18–0.26	—
Blowing agent	23.0 CFCl$_3$	13.65 CFCl$_3$
Combustion modifier (flame retardant)	13.0[c]	4–5[d]

[a] Data from Ref. [11].
[b] The isocyanate concentration is calculated at 90–120% of the theoretical amount required to react with both hydroxyl groups and water. This percentage is commonly known as the isocyanate index.
[c] Chlorinated phosphate ester.
[d] Usually a chlorinated or brominated compound—some may have reactive hydroxyl groups and additional isocyanate would have to be added.

8 POLYURETHANES

the triols. The use of aromatic diamines leads to a polyurea of greater rigidity than a polyether polyurethane. Some typical flexible and rigid polyurethane foam formulations are shown in Table I.

A good source of up-to-date information is from the major foam component suppliers such as Mobay, Union Carbide, ICI, Air Products, Elf Atochem North America, Du Pont, BASF, and Ethyl.

Refrigerant IT or CFC-11 ($CFCl_3$) blowing agent is slowly being replaced by other blowing agents such as HCFC-141b (CH_3CCl_2F), which are thought to be safer for the environment [15].

8-1. Preparation of a Polyurethane Prepolymer [16]

$$\begin{array}{c} CH_3 \\ | \\ CH-CH_2 \\ | \\ OH \end{array} \left[\begin{array}{c} CH_3 \\ | \\ CH-CH_2O \\ \end{array} \right]_n CH_2 - \begin{array}{c} CH_3 \\ | \\ CH \\ | \\ OH \end{array} + 2 \underset{NCO}{\underset{|}{\bigodot}}\overset{CH_3}{\underset{}{}}-NCO \longrightarrow$$

$$\begin{array}{c} CH_3 \\ | \\ CH-CH_2 \\ | \\ OCNH \\ \parallel \\ O \\ | \\ \underset{CH_3}{\bigodot}-NCO \end{array} \left[\begin{array}{c} CH_3 \\ | \\ CH-CH_2O \\ \end{array} \right]_n CH_2 - \begin{array}{c} CH_3 \\ | \\ CH \\ | \\ OC-NH \\ \parallel \\ O \\ | \\ \underset{CH_3}{\bigodot}-NCO \end{array} \qquad (2)$$

To a 500 ml, dry, nitrogen-flushed resin kettle equipped with a mechanical stirrer, thermometer, condenser with drying tube, dropping funnel, and heating mantle are added 34.8 gm (0.20 mole) of 2,4-toluene diisocyanate and 0.12 gm (0.000685 mole) of o-chlorobenzoyl chloride. Polyoxypropylene glycol (Dow Chemical Co. P-1000, mol. wt. 1000) is added dropwise over a 2 hr period at 65° ± 3°C until 100.0 gm (0.1 mole) has been added. Stirring at 65°C is maintained for an additional 2 hr and then the reaction is cooled. The final product is a clear viscous liquid; NCO calc. 6.23%; found 6.27%. Several other preparations of polyurethane prepolymers have been reported [17].

Various catalysts help to speed the reaction of isocyanate groups with alcohols.

Tertiary amines activate the isocyanate functional groups for reaction with an active hydrogen compound [18].

8-2. Preparation of Poly[ethylene methylene bis(4-phenylcarbamate)] [19, 20]

$$HOCH_2CH_2OH + CH_2\left(\underset{}{\bigcirc}-NCO\right)_2 \longrightarrow$$

$$\left[-\overset{O}{\underset{\|}{C}}NH-\underset{}{\bigcirc}-CH_2-\underset{}{\bigcirc}-NH-\overset{O}{\underset{\|}{C}}-OCH_2CH_2O-\right]_n \quad (3)$$

To a flask occupied with a mechanical stirrer, condenser, and dropping funnel and containing 40 ml of 4-methylpentanone-2 and 25.02 gm (0.10 mole) of methylene bis(4-phenyl isocyanate) (MDI) is added all at once 6.2 gm (0.10 mole) of ethylene glycol in 40 ml of dimethyl sulfoxide. The reaction mixture is heated at 115°C for 1.5 hr, cooled, poured into water, and filtered. The white polymer is chopped up in a blender, washed with water, and dried under reduced pressure at 90°C to afford 29.6–31.2 gm (95–100%), $\eta_i =$ 1.01 (0.05% solution in DMF at 30°C), polymer melt temperature, 240°C.

REFERENCES

1. S. R. Sandler and W. Karo, "Organic Functional Group Preparations," Vol. 2, 2nd ed., Academic Press, New York, 1986.
2. O. Bayer, *Angew. Chem.* **59**, 275 (1947).
3. O. Bayer, H. Rinke, W. Siefken, L. Orthner, and H. Schild, German Patent 728,981 (1942).
4. H. Rinke, H. Schild, and W. Siefken, French Patent 845,917 (1939); U.S. Patent 2,511,544 (1950).
5. E. L. Wittbecker and P. W. Morgan, *J. Polym. Sci.* **40**, 289 (1959).
6. E. L. Wittbecker and M. Katz, *J. Polym. Sci.* **40**, 367 (1959).
7. W. E. Catalin, U.S. Patent 2,284,637 (1942); W. E. Hanford and D. F. Holmes, U.S. Patent 2,284,896 (1942); R. E. Christ and W. E. Hanford, U.S. Patent 2,333,639 (1943); F. B. Hill, C. A. Young, J. A. Nelson, and R. G. Arnold, *Ind. Eng. Chem.* **48**, 927 (1956); H. Rinke, *Chimia* **16**, 93 (1962); *Angew. Chem., Int. Ed. Engl.* **1**, 419 (1962); O. Bayer, E. Müller, S. Petersen, H. F. Piepenbrink, and E. Windemuth, *Angew. Chem.* **62**, 57 (1950); C. S. Marvel and J. H. Johnson, *J. Amer. Chem. Soc.* **72**, 1674 (1950); E. Dyer and G. W. Bartels, *ibid.* **76**, 591 (1954).

8. Y. Miyake, S. Ozaki, and Y. Hirata, *J. Polym. Sci., Part A-1* **7**, 899 (1969).
9. *Polyurethanes 90 Proceedings of the SPI 33rd Annual Technical Marketing Conference*, September 30–October 3, 1990, Orlando, Florida. Technomic Publishing Co., Inc., Lancaster, Pennsylvania, 1990.
10. "Threshold Limit Values and Biological Exposure Indices," American Conference of Governmental Industrial Hygienists (ACGIH), Cincinnati, Ohio, 1986–1987.
11. J. K. Backus, C. D. Blue, P. M. Boyd, F. J. Caine, J. H. Chapmen, J. L. Eakin, S. J. Harasin, E. R. McAfee, C. G. McCarty, N. H. Nodelman, J. H. Rieck, H. G. Schmelzar, and E. P. Squilter, "Polyurethanes," *Encycl. Polym. Sci. Eng.* **13**, 243 (1988).
12. G. E. Burrows, *Cell. Polym.* **2** (3), 205 (1983).
13. H. H. G. Jellinek and S. R. Dunkle, *J. Polym. Sci., Polym. Ed.* **21**, 487 (1983).
14. J. H. Saunders and K. C. Frisch, "Polyurethanes—Chemistry and Technology," Wiley (Interscience), New York, 1964.
15. J. P. Lavelle and T. Doyle *in* Polyurethanes 90 [9], pp. 407–412.
16. S. R. Sandler and F. R. Berg, *J. Appl. Polym. Sci.* **9**, 3909 (1965).
17. S. R. Sandler, *J. Appl. Polym. Sci.* **11**, 811 (1967).
18. J. W. Baker and J. B. Holdsworth, *J. Chem. Soc., London* p. 713 (1947); J. W. Baker, J. Gaunt, and M. M. Davies, *ibid.*, p. 9 (1949); A. Farkas and K. G. Flynn, *J. Amer. Chem. Soc* **82**, 642 (1950); K. G. Flynn and D. R. Nenortas, *J. Org. Chem.* **28**, 3527 (1963).
19. D. J. Lyman, *J. Polym. Sci.* **45**, 49 (1960).
20. H. C. Beachell and J. C. Peterson, *J. Polym. Sci., Part A-1* **7**, 2021 (1969).

9
THERMALLY STABLE POLYMERS

The search for thermally stable resins usable at 500°C for long periods of time has been given great support as a result of the material needs of space and advanced aircraft industries.

Thermally stable polymers should have the following properties: (1) high melting or softening points; (2) low weight loss as determined by thermogravimetric analysis; (3) structures that are not susceptible to degradative chain scission or intra- or intermolecular bond formation; and (4) chemically inert especially to oxygen, moisture, and dilute acids and bases.

Chemical structures which are thermally stable usually also have (1) a highly resonance-stabilized system; (2) an aromatic or other thermally unreactive ring structure as a major portion of the polymer composition; and (3) a high bond and cohesive energy density.

In view of these requirements, the most promising thermally stable polymers are the following: polyimides [1, 2], polybenzimidazoles [1–6], polyoxadiazoles [4, 7], polytriazoles [7, 8], polybenzoxazoles [4, 8], polyazomethines [9], polyphenylenes [10, 11], polytriazines [12], polyquinoxalines [13–15], polythiazoles [8], polyphenylene ethers [16–20], polyaryl sulfones [2, 21], and related polymeric sulfur analogs. More recently, polyketone type polymers have become available which can be used at 500°F for several thousand hours. Various modifications include polyether ketones, polyetherether ketones, etc. A typical polymer is described in Ref. [22] and the thermal behavior in Ref. [23]. Many others are constantly being reported. Space limitations will not permit them all to be described in this chapter.

Most of these polymers are prepared by condensation polymerization techniques, with the exception of the polyphenylene ethers, which probably are prepared via a radical mechanism as described below.

Polymers which possess the desirable thermal properties are usually difficult to fabricate into products. Recently prepolymers which are workable before cyclization have been employed to extend the practical applications of these polymers.

Notable among polymers utilizing these prepolymers have been the polypyromellitimides [2] and the polybenzimidazoles [3].

Several worthwhile reviews [24-27] have appeared and should be consulted for additional information. The status of fibers useful at high temperatures has been reviewed [28].

- **CAUTION:** Several high-temperature polymer starting materials such as benzidine, 3,3′-diaminobenzidene, and other aromatic amines should be handled with care since they are known to be carcinogenic [29].

Polycarbonates have already been described in Chapter 3 and will not be discussed further here.

Polyphenylene ethers have rapidly developed into an important commercial engineering thermoplastic since the discovery of the oxidative coupling of phenols (see below). Noryl® (General Electric) is a typical commercial resin comprising alloys of poly(2,6-dimethyl-1,4-phenylene ether) and high impact polystyrene. The polymer poly(2,6-dimethyl-1,4-phenylene ether) is also produced by General Electric under the trademark PPO. The Noryl resins can also be flame retarded and are processed by injection molding, foam molding, extrusion thermoforming, or blow molding, for use in a variety of applications (automotive, electrical, construction, appliance, business machines, etc.). For further information a recent review of this subject is recommended [30].

Hunter [30-37] was the first to prepare a poly(phenylene oxide) by the ethyl iodide-induced decomposition of anhydrous silver 2,4,6-tribromo-phenoxide [32, 33]. The polymer was isolated as a white, amorphous powder with a softening point above 300°C. The molecular weight was estimated to be about 6,000-12,000. Hunter [35] reported in a series of papers the polymerization of other trihalophenols and established that iodine was displaced more readily than bromine and the latter more rapidly than chlorine. The

p-halogen reacted more rapidly than the o-chlorine group [36] and only halogen with a free ionizable phenolic group could be displaced [37].

- **CAUTION:** All phenols discussed below are toxic materials and must be handled with caution.

9-1. Preparation of Poly(2,6-dimethyl-1,4-phenylene oxide) by Polymerization of Silver 4-Bromo-2,6-dimethylphenolate [38]

$$\text{Br}\underset{CH_3}{\overset{CH_3}{\bigcirc}}\text{OH} \longrightarrow \text{Br}\underset{CH_3}{\overset{CH_3}{\bigcirc}}\text{OAg} \longrightarrow$$

$$\text{Br}\underset{CH_3}{\overset{CH_3}{\bigcirc}}\text{O}\left[\underset{CH_3}{\overset{CH_3}{\bigcirc}}\text{O}\right]_n\underset{CH_3}{\overset{CH_3}{\bigcirc}}\text{OH} + \text{AgBr} \quad (1)$$

(a) Preparation of silver salt [31, 38]. To an Erlenmeyer flask containing 100 ml of distilled water is added 20.1 gm (0.1 mole) of 4-bromo-2-6-dimethylphenol and 5.6 gm (0.1 mole) of potassium hydroxide dissolved in 100 ml of distilled water. The resulting solution is filtered and to the filtrate 1% acetic acid is added dropwise to give a faint turbidity. A few drops of dilute silver nitrate are then added and any resulting silver bromide filtered. Then to the filtrate is added a solution of 17.0 gm (0.1 mole) of silver nitrate in 100 ml of water to give a voluminous precipitate of the silver salt. After storage in the dark for 1–2 hr the precipitate is filtered, washed with water, and dried under reduced pressure to give an almost quantitative yield of the silver salt.

(b) Polymerization of silver salt [38]. To a flask is added 4.0 gm (0.013 mole) of silver 4-bromo-2,6-dimethylphenolate and 100 ml of benzene. The suspension is refluxed for 1 hr to give a precipitate of silver bromide. The benzene solution is added to 300 ml of methanol to precipitate the polymer. The polymer is dried under reduced pressure to afford 0.79 gm (50%) with a softening point of 210°–220°C and $[\eta] = 0.06$ dl/gm in $CHCl_3$ at 25°C. The mol. wt.

of 1600 is determined by an ebuliometric method in benzene or chloroform.

Price [19] reinvestigated this problem and studied the polymerization of 4-bromo-2,6-dimethylphenol with several oxidizing agents such as ferricyanide in PbO_2, I_2, O_2, and light. The soluble polymer had molecular weights in the range 2,000–10,000. Sterically hindered halogenated phenols (4-bromo-2,6-di-*tert*-butylphenol, bromodurenol, and pentabromophenol) failed to polymerize under the same conditions. A radical mechanism (see Eq. 2) was suggested for this reaction, and was later confirmed.

$$(2)$$

Hay [16, 39] and co-workers were the first to report that oxidative coupling of 2,6-disubstituted phenols gave high molecular weight 2,6-disubstituted 1,4-phenylene ethers (for $R = CH_3$). When the substituents are bulky, such as $R = $ *tert*-butyl, then only carbon–carbon coupling occurs to give diphenoquinones as the sole product. When $R = $ isopropyl then both reactions shown in Eqs. (3) and (4) occur.

In order to overcome the steric reduction in rate for C—O coupling with 2,6-diphenylphenol, the reaction was carried out at 60°C using a Cu–N,N,N',N'-tetramethyl-1,3-butanediamine catalyst to give polymers in high yield having a molecular weight ($\overline{M}_n$) of approx. 300,000 [40, 41]. Only 3% of the diphenoquinone was formed [40, 41]. The polymerization of various other 2,6-diarylphenols has also been reported [42] (see Eqs. 3 and 4).

$$n \underset{R}{\underset{R}{\bigcirc}}\text{-OH} + \tfrac{1}{2}n\text{O}_2 \xrightarrow{\text{Cu-amine}} \left[\underset{R}{\underset{R}{\bigcirc}}\text{-O}\right]_n + n\text{H}_2\text{O} \quad (3)$$

$$\downarrow \tfrac{1}{2}n\text{O}_2$$

$$\tfrac{1}{2}n\text{HO}\underset{R}{\underset{R}{\bigcirc}}\underset{R}{\underset{R}{\bigcirc}}\text{OH} \xrightarrow{\tfrac{1}{2}n\text{O}_2} \tfrac{1}{2}n\text{O}\underset{R}{\underset{R}{\bigcirc}}\underset{R}{\underset{R}{\bigcirc}}\text{O} + n\text{H}_2\text{O} \quad (4)$$

Hay [42] reported that the oxidative polymerization of 2,6-diphenylphenol could be effected with either oxygen–amine copper salts [41], lead dioxide [43], or silver oxide [44].

Block copolymers of polyphenylene ethers starting with 2,6-dimethylphenol and 2,6-diphenylphenol were prepared using a catalyst system consisting of cuprous bromide, N,N,N',N'-tetramethyl-1,3-butanediamine, and magnesium sulfate [18]. Recently NMR data have been presented to show that a random copolymer is produced using a cuprous chloride-pyridine catalyst at 25°C [20, 45].

$$\left[\underset{\text{CH}_3}{\underset{\text{CH}_3}{\bigcirc}}\text{-O}\right]_x \left[\underset{\text{C}_6\text{H}_5}{\underset{\text{C}_6\text{H}_5}{\bigcirc}}\text{-O}\right]_y$$

9-2. Preparation of Poly(2,6-dimethyl-1,4-phenylene oxide) Using CuCl–Pyridine–O_2 [16, 46]

$$n\underset{\text{CH}_3}{\underset{\text{CH}_3}{\bigcirc}}\text{-OH} + \tfrac{1}{2}n\text{O}_2 \xrightarrow[\text{C}_6\text{H}_5-\text{NO}_2]{\text{CuCl} \atop \text{pyridine}} \left[\underset{\text{CH}_3}{\underset{\text{CH}_3}{\bigcirc}}\text{-O}\right]_n + n\text{H}_2\text{O} \quad (5)$$

To a 50 ml flask connected to an oxygen buret (an oxygen addition tube can also be used in place of a buret) is added 15 ml of nitrobenzene, 4.5 ml of pyridine, and 0.020 gm (0.0002 mole) of finely divided cuprous chloride. The mixture is stirred rapidly for 15 min to give a dark green solution. At this point 0.49 gm (0.004 mole) of 2,6-xylenol is added and stirring is continued. After 20 min the theoretical amount of oxygen uptake is complete and the orange reaction mixture is stirred for an additional 20 min (total oxygen uptake, 101.5%). At this time the orange color is almost gone and the solution is added dropwise to 65 ml of methanol acidified with 0.65 ml of conc. HCl. The precipitated polymer is filtered and washed with additional acidified methanol. The polymer is reprecipitated twice from solution in 10 ml of benzene by adding to 40 ml of cold methanol. Finally the polymer is dissolved in 20 ml of benzene, washed with water, and freeze-dried at 1 mm Hg pressure to afford 0.40 gm (84%) of polymer, $\eta_{intrinsic}$ = 0.95 to 1.0 dl/gm; osmotic molecular weight, 28,000; and softening point, 240°C. The infrared spectrum of the polymer shows no absorption due to OH groups.

The same polymer is obtained when one starts with 2,6-dimethyl-4-halophenol (halogen = Br or Cl) but equimolar amounts of cuprous chloride are required.

The linear polyphenylene ethers from the amine–copper-catalyzed polymerization have a DP greater than 200, with softening points above 240°C [47].

The polymerization reaction has been shown to involve a free-radical step-growth process. The final polymer may be prepared from either the monomer, dimer, trimer, or other isolated low molecular weight intermediates [48]. The oxygen has the function of keeping the copper ions in the active divalent state [38, 49].

The use of excess amine in the amine–Cu complex gives a more active catalyst with higher rate yields of C—O products rather than those formed by C—C bonding. This evidence supports the suggestion that the catalyst using pyridine probably can be represented by Eq. (6):

$$Cu^{I}Pyr + Pyr \xrightarrow{O_2} Cu^{II}Pyr + Pyr \rightleftharpoons Cu^{II}Pyr - Pyr \quad (6)$$

REFERENCES

1. R. Phillips and W. W. Wright, *J. Polym. Sci., Part B* **2**, 47 (1964).
2. C. E. Sroog, A. L. Endrey, S. V. Abramo, C. E. Berr, W. M. Edwards, and K. L. Olivier, *J. Polym. Sci., Part A* **3**, 1373 (1965).

3. K. C. Brinker and I. M. Robinson, U.S. Patent 2,895,984 (1959).
4. N. M. Koton, *Russ. Chem. Rev.* **31**, 81 (1962).
5. H. Vogel and C. S. Marvel, *J. Polym. Sci.* **50**, 511 (1961).
6. H. Vogel and C. S. Marvel, *J. Polym. Sci., Part A* **1**, 1531 (1963).
7. P. M. Hergenrother, *Macromolecules* **3**, 10 (1970).
8. V. V. Korshak and M. M. Teplyakov, *J. Macromol. Sci., Rev. Macromol. Chem.* **5**, 409 (1971).
9. G. F. D'Aleilo and R. E. Schoenig, *J. Macromol. Sci., Rev. Macromol. Chem.* **3**, 108 (1969).
10. C. S. Marvel and G. E. Hartzell, *J. Amer. Chem. Soc.* **81**, 488 (1959); C. S. Marvel and N. Tarkoy, *ibid.* **79**, 6000 (1957); J. G. Speight, P. Kovacic, and F. W. Koch, *J. Macromol. Sci., Rev. Macromol. Chem.* **5**, 295 (1971).
11. J. K. Stille and G. K. Noren, *Macromolecules* **5**, 49 (1972).
12. W. Wrasidlo and P. M. Hergenrother, *Macromolecules* **3**, 548 (1970).
13. P. M. Hergenrother, *J. Macromol. Sci., Rev. Macromol. Chem.* **6**, 1 (1971).
14. W. Wrasidlo and J. M. Augl, *Macromolecules* **3**, 544 (1970).
15. P. M. Hergenrother and D. E. Kiyohara, *Macromolecules* **3**, 387 (1970).
16. A. S. Hay, H. S. Blanchard, G. F. Endres, and J. W. Eustance, *J. Amer. Chem. Soc.* **81**, 6335 (1959).
17. G. D. Cooper, H. S. Blanchard, G. F. Endres, and H. L. Finkbeiner, *J. Amer. Chem. Soc.* **87**, 3996 (1965).
18. J. G. Bennett, Jr. and G. D. Cooper, *Macromolecules* **2**, 101 (1969).
19. G. D. Staffin and C. C. Price, *J. Amer. Chem. Soc.* **82**, 3632 (1960).
20. G. D. Copper and J. G. Bennett, Jr., *J. Org. Chem.* **37**, 441 (1972).
21. *Chem. & Eng. News* **43**, 48 (1965); Union Carbide, Netherlands Patent Appl. 6,604,731 (1966).
22. K. J. Dahl, U.S. Patent 4,111,908 (1978).
23. H. Gupta and R. Salova, *Polym. Eng. Sci.* **30**, (8), 453 (1990); L. M. Robeson, P. A. Winslow, M. Matzner, J. E. Harris, and L. M. Marcesca, U.S. Patent 4,908,425 (1990).
24. C. L. Segal, ed., "High Temperature Polymers," Dekker, New York, 1967; J. E. Mulvaney, *Encycl. Polym. Sci. Technol.* **7**, 478 (1967).
25. M. S. Reisch, *Chem. Eng. News* **67** (36), 21 (1989).
26. P. M. Hergenrother, *Chem. Tech.* **14** (8), 496 (1984).
27. P. M. Hergenrother, *Encycl. Poly. Sci. Eng.* **7**, 639 (1987).
28. J. Preston, *Chem. Technol.* p. 664 (1971).
29. C. E. Searle, *Chem. & Ind.* (*London*), p. 111 (1972); "The Carcinogenic Substances Regulations," No. 879, HM Stationary Office, London, 1967.
30. D. Aycock, V. Abolins, and D. M. White, *Encycl. Polym. Sci. Eng.* **13**, 1 (1988) (2nd ed.).
31. H. W. Hunter, A. Olson, and E. A. Daniels, *J. Amer. Chem. Soc.* **38**, 1761 (1916).
32. A. Hantzsch, *Ber. Deut. Chem. Ges.* **40**, 4875 (1907).
33. H. A. Torrey and W. H. Hunter, *J. Amer. Chem. Soc.* **33**, 194 (1911).
34. W. H. Hunter and G. H. Woollett, *J. Amer. Chem. Soc.* **43**, 131, 135, and 1761 (1921).
35. W. H. Hunter and M. A. Dahlen, *J. Amer. Chem. Soc.* **54**, 2459 (1932).

36. W. H. Hunter and F. E. Joyce, *J. Amer. Chem. Soc.* **39**, 2640 (1917).
37. W. H. Hunter and M. A. Dahlen, *J. Amer. Chem. Soc.* **54**, 2456 (1932).
38. H. S. Blanchard, H. L. Finkbeiner, and G. A. Russell, *J. Polym. Sci.* **58**, 469 (1962).
39. A. S. Hay, French Patents 1,322,152 (1963); 1,384,255 (1965); Belgian Patents 635,349 and 635,350 (1964); U.S. Patents 3,262,892 (1966); 3,432,466 (1969).
40. A. S. Hay and D. M. White, *Polym. Prepr., Amer. Chem. Soc., Div. Polym. Chem.* **10**, 92 (1969).
41. A. S. Hay, *Macromolecules* **2**, 107 (1969).
42. A. S. Hay and R. F. Clark, *Macromolecules* **3**, 533 (1970).
43. H. M. V. Dort and C. R. H. I. de Jonge, U.S. Patent 3,400,100 (1968).
44. B. O. Lindgren, *Acta Chem. Scand.* **14**, 1203 (1960).
45. G. D. Cooper and J. G. Bennett, Jr., *Polym. Prepr., Amer. Chem. Soc., Div. Polym. Chem.* **13**, 551 (1972).
46. General Electric Co., British Patent 930,993 (1963); W. A. Butte, Jr. and C. C. Price, *J. Amer. Chem. Soc.* **84**, 3567 (1962).
47. A. S. Hay, *Polym. Prepr., Amer. Chem. Soc., Div. Polym. Chem.* **2**, 319 68 (1961); G. F. Endres and J. Kwiatek, *J. Polym. Sci.* **58**, 593 (1962).
48. C. C. Price and N. S. Chu, *J. Polym. Sci.* **61**, 135 (1962).
49. C. R. G. R. Bacon and H. A. O. Hill, *Quart. Rev., Chem. Soc.* **19**, 114 (1965); E. Ochiai, *Tetrahedron* **20**, 1831 (1964); A. I. Scott, *Quart. Rev., Chem. Soc.* **19**, 2 (1965).

10
POLYMERIZATION OF ACRYLIC MONOMERS

The polymerization procedures of monomers related to acrylic acid (e.g., acrylic, methacrylic, itaconic acids, and their salts; acrylonitrile; acrylamide; and the esters of acrylic and methacrylic acids) are similar. Of course, due allowances for their individual characteristics must be made. Among the properties that give rise to the individuality of each class of monomer are physical state, solubility in various solvents, solubility of the polymer in its monomer, the nature of the inhibitors, the ease of complex formation, the copolymerization reactivity ratios, etc. Thus, for example, all of the liquid monomers may be polymerized in bulk. Solids such as itaconic acid or acrylamide can be bulk polymerized by radiation techniques and, presumably, also as melts.

All of these monomers may be polymerized in some suitable solvents. Those monomers which are only moderately soluble in water can be polymerized by suspension and emulsion techniques.

In this chapter, for each group of monomers, those techniques will be emphasized that are of particular interest for each particular class of starting materials. We should mention that this does not mean that the procedures given herein are the only ones that are suitable.

Based on S. R. Sandler and W. Karo, "Polymer Syntheses," 2nd ed., Vol I, pp. 318ff, 378ff, 420ff, and Vol. 2, pp. 300ff, by permission of Academic Press, Inc., San Diego, 1994.

1. POLYMERIZATION OF ACRYLIC ACIDS AND RELATED COMPOUNDS

Particularly before acrylic acid became commercially available, poly(acrylic acid), various salts of the acid, and its copolymers were prepared by hydrolysis of poly(acrylonitrile) or appropriate copolymers of acrylonitrile. Polymeric esters of acrylic esters also have been hydrolyzed to poly(acrylic acid). Poly(methacrylate esters) may be hydrolyzed to poly(methacrylic acid); however, the saponification of poly(methacrylates) usually requires more drastic conditions than that needed for the poly(acrylates). In general, the hydrolytic procedures may lead to incomplete reaction as well as to chain degradation. Even so, it is thought that the hydrolysis of well-defined poly(acrylic esters) retains the characteristics of the backbone such as its tacticity.

10-1. Preparation of Isotactic Poly(acrylic acid) [1]

To a hot solution of 1 gm (0.009 mole) of isotactic poly(isopropyl acrylate) in 5 ml of toluene is added 10 ml of a solution containing 0.070 gm of potassium hydroxide in 1 ml of isopropanol. The mixture is refluxed for 6 hr at 96°–97°C. Then the reflux condenser is set down for distillation, water is added to the reaction mixture, and the water–isopropanol is slowly distilled off while the isopropanol is replaced with water. During this step, which is said to take 10 hr, the polymer goes completely into solution. The solution is cooled, and an excess of methanol is added to precipitate potassium poly(acrylate). The polymer is filtered off, washed with methanol, and dried.

The polymer is dissolved in a small quantity of water, and an excess of concentrated hydrochloric acid is added to precipitate poly(acrylic acid). The polymer is filtered off, washed with concentrated hydrochloric acid, and dried under reduced pressure at room temperature. By titration this polymer appears to be only 84.7–85% hydrolyzed, but X-ray diffraction shows that crystalline species have been isolated. On drying to constant weight at 80°C, titration indicates 89.6% hydrolysis.

A. BULK POLYMERIZATION (THERMALLY OR CHEMICALLY INITIATED) [2]

The heat of polymerization of acrylic acid is 18.5 kcal/mole and that of methacrylic acid 15.7 kcal/mole [3]. Perhaps, when these values are converted to 260 cal/gm for acrylic and 180 cal/gm for methacrylic acid, the hazard involved in bulk polymerizations of these monomers may be appreciated more readily. For this reason, only very small-scale bulk polymerizations should be attempted, with suitable protective measures.

Thermal initiation of acrylic acid polymerization has been reported to be very rapid at elevated temperatures. The resultant product is said to be partially insoluble as well as being partially degraded [3].

The bulk polymerization of glacial methacrylic acid has been carried out by heating 5 gm of the acid with 0.1 gm of 2,2'-azobis(isobutyronitrile) at 60°C for 0.5 hr. The resultant white mass was soluble in methanol, dioxane, and tetrahydrofuran [4].

Acrylic and methacrylic acids are common constituents in copolymer systems. If the copolymerizations are carried out in "inert" solvents, there are variations in reactivity ratios that are related to the solvent. One would also expect pH effects if the processes were carried out in aqueous media. With comonomers that have limited solubilities in water, complications related to the distribution coefficients come into play. Some aspects of these situations in the copolymerization of acrylonitrile with methacrylic acid are reviewed in Ref. [5].

B. SUSPENSION POLYMERIZATION

The general principle of suspension polymerization has been discussed in Sandler and Karo [6]. Since both acrylic and methacrylic acids are quite soluble in water, the ordinary suspension polymerization procedures are not applicable for conversion of these monomers to their homopolymers (although they may be applicable to copolymerization systems). If the aqueous phase contains a fairly high proportion of dissolved electrolytes, the monomeric acids are "salted out." Then they may be dispersed by agitation and suspension polymerization procedures applied. The resulting product is water insoluble but swellable with water. In aqueous solution of the alkaline hydroxides or of ammonia, these polymers are readily soluble.

The method is said to have the advantage over aqueous solution polymerizations in producing products of high molecular weight quite rapidly and in permitting the usual controls associated with suspension polymerization techniques.

Suspension copolymers of acrylic acids with many comonomers have been prepared. In some cases the acids are present in only small amounts to modify the bulk of the resin beads; in other cases a major fraction of the resin may be an acrylic acid. By the proper balancing of the copolymer composition, materials have been prepared that may permit sustained releases of pharmaceuticals. Chloramphenicol or chlorothiazide have been incorporated in bead polymers as possible delayed-action or sustained-release preparations [7].

10-2. Preparation of Poly(methacrylic acid) by Suspension Polymerization with a Monomer-Soluble Initiator [8]

In a flask equipped with an explosion-proof mechanical stirrer with adjustable speed, an additional funnel, a thermometer, and a reflux condenser, 80 gm of anhydrous sodium sulfate is dissolved in 400 gm of water. Then 8 gm of a 10% aqueous solution of sodium poly(acrylate) is added as the suspending agent. After the addition of 100 gm (1.16 moles) of inhibited methacrylic acid containing 0.2 gm (0.0008 mole) of benzoyl peroxide, the reaction system is stirred continuously but not too rapidly and heated at 80°C with a water bath for approximately 2.5 hr. A small quantity of the 10% aqueous solution of sodium poly(acrylate) is maintained in the addition funnel throughout the process so that part of this reagent may be added if there are indications of lump formation.

After the initial heating period, the reaction temperature is raised to 90°–95°C for a short time. The solid product is collected on a nylon chiffon cloth and washed free of the sodium sulfate and the suspending agent with a 5% aqueous solution of hydrochloric acid. Then the product is dried under reduced pressure. The yield is essentially quantitative. The product is said to be a sandy mass, virtually ash-free, and soluble in alkali hydroxides to produce highly viscous solutions.

C. SOLUTION POLYMERIZATION

Monomeric acrylic and methacrylic acids are soluble in a large variety of organic solvents. However, the polymers derived from these monomers often exhibit limited solubility in these solvents. Frequently their solubility increases at low temperatures.

A procedure that yields substantial laboratory quantities of poly(methacrylic acid) of high molecular weight (e.g., 0.84–1.7 × 10^6 as determined by intrinsic viscosity measurements) is given in Procedure 10-3.

10-3. Polymerization of Methacrylic Acid in an Aqueous Solution [8, 9]

In a suitable reaction vessel, a solution of 180 gm (2.1 moles) of freshly distilled methacrylic acid in 900 ml of distilled water is sparged with oxygen-free nitrogen. Then 5.4 gm (0.05 mole) of a 30% hydrogen peroxide solution is added. The mixture is maintained between 80° and 90°C for 5 hr. The gelled polymer system is allowed to cool to room temperature. The polymer is dissolved in distilled methanol and precipitated with diethyl ether. The resulting polymer is air dried, powdered, and then dried under reduced pressure. Yield: 178 gm (99%) of atactic poly(methacrylic acid).

NOTE: During the preparation and during polymer fractionation, it is important to avoid contact of the poly(methacrylic acid) solution with metallic objects.

The purification of 20% aqueous poly(methacrylic acid) solutions by dialysis against distilled water, followed by concentration of the solutions in an evaporator to a 2% water content and final drying, has been reported. The recovered polymer should be stored in sealed ampoules in the cold since there is said to be a slow degradation at room temperature [10].

D. CHARGE-TRANSFER COMPLEX AND OTHER COMPLEX POLYMERIZATIONS

A study of the copolymerization of acrylic acid and acrylamide in the presence of zinc chloride and 2,2'-azobis(isobutyronitrile) (AIBN) in benzene showed that alternating copolymers form. Since copolymers of these two monomers prepared by ordinary free-radical methods normally produce random copolymers, it is presumed that charge-transfer complexes are involved when zinc chloride is present in the system [11].

In benzene solution, the system is heterogeneous. When no AIBN is present, no polymer forms. In the absence of benzene, the acrylamide-acrylic acid-zinc chloride compositions form homogeneous compositions. However, even after two months at room temperature, no evidence of polymerization was noted. To form polymers, AIBN and a temperature of 45°C are necessary.

10-4. Charge-Transfer Complex Polymerization of Acrylic Acid–Acrylamide [11]

In a flask fitted with a mechanical stirrer, condenser, thermometer, and nitrogen inlet tube, a solution of the monomers given in Table I is dissolved in 50 ml of benzene and sparged with nitrogen. While the solution is stirred, the requisite quantity of zinc chloride and 300 mg of AIBN is added. Then the reaction flask is placed in a thermostatted water bath at 45° ± 1°C.

After the desired reaction time, the polymerization is stopped by cooling to 0°C. The copolymer is precipitated by addition to a solution of 300 ml methanol, 10 ml of water, and 1 ml of concentrated hydrochloric acid. The copolymers are purified by dissolving them in water and reprecipitating with methanol.

Reaction conditions for the copolymerization of acrylic acid with acrylamide with several variation in the level of zinc chloride are given in Table I. From the data presented there, it appears that the molecular weight of the polymer increases with increasing zinc chloride concentration, although conversion seems to have been reduced under the reaction conditions. Increasing the level of acrylic acid over acrylamide led to a reduction in molecular weight, whereas, with a higher concentration of acrylamide in the monomer charge, the molecular weight rose.

The chemistry of template and other complex-forming polymerizations is of great potential interest. For example, Higashi, Nojima, and Niwa point out that the interaction of two different macromolecules is thought to be important in living systems. Enzyme processes and the recognition of molecules on a biomembrane surface are closely related to this aspect of chemistry. Their work dealt with the recognition of poly(acrylic acid) by poly(methacrylamide)-bearing terminal stearyl groups [12].

TABLE I
Copolymerization of Acrylic Acid and Acrylamide in the Presence of Zinc Chloride[a,b]

Monomer charge			Reaction temp. (°C)	Reaction time (min)	Conversion (%)	Percent acrylic acid in polymer	$\overline{M}_v \times 10^{-5}$ [c]
Acrylic acid (moles)	Acrylamide (moles)	Zinc chloride (moles)					
0.10	0.10	—	45	5	20.6	66.2	1.55
0.10	0.10	0.10	45	5–7	61.9	50.5	1.20
0.10	0.10	0.50	45	5–6	10	47.0	4.79
0.15	0.10	0.50	49	5–6	14.5	53.7	3.89
0.10	0.15	0.50	55	8	19.9	44.2	6.30
0.05	0.45	0.50	—	—	—	47.4	17.4

[a] From Ref. [10].
[b] Solvent: 50 ml of benzene; initiator 300 mg of AIBN.
[c] Viscosities were determined in 1 M sodium nitrate at 30°C; the relationship of intrinsic viscosity and molecular weight, $\overline{M}_v$, used was $[\eta] = 3.73 \times 10^{-4}\, \overline{M}_v^{0.66}$.

E. POLYMERIZATION OF ITACONIC ACID

The polymerization of itaconic acid seems not to have been studied very extensively, although industrial applications in copolymer systems appear to be of considerable interest. In view of the work of A. Katchalsky and coworkers [13] on the effect of pH on the polymerization of acrylic acid and methacrylic acid, anlogous research on pH effects on itaconic acid reactions has been carried out to a limited extent [14, 15]. Typically, with a persulfate initiator in an aqueous solution at 50°C, the monomer is converted to an extent of 85–90% to its homopolymer within 35–45 hr. In the pH range of 2.3–3.8, the rate of polymerization is constant. As the pH increases, the rate becomes progressively slower and stops completely at a pH of 9. Generally, the last 5–10% of the monomer seems to be difficult to convert to polymer.

Preparation 10-5 is one of the early examples of the polymerization in a strongly acid system. This polymer evidently seemed to contain only 40% of the expected carboxylic acid groups, and its elementary analysis deviated considerably from calculated values [15].

10-5. Bottle Polymerization of Itaconic Acid [15]

In a 4 oz. screw-cap bottle, sealable with a Buna-N gasket, are placed 50 ml of 0.5 N hydrochloric acid (prepared by diluting concentrated hydrochloric acid with distilled water that had been deaerated by heating to vigorous boiling and cooling while bubbling oxygen-free nitrogen through it), 20 gm ($\sim$ 0.15 mole) of itaconic acid, and 0.10 gm (0.00036 mole) of potassium persulfate.

The bottle is flushed with oxygen-free nitrogen to remove air, sealed, covered with a protective steel jacket, and placed in a thermostatted bath at 50°C for 68 hr. During this treatment some of the itaconic acid remains undissolved.

The bottle is cooled to room temperature. The solution is slowly added to well-stirred acetone to precipitate the polymer. The product is filtered off, redissolved in water, and reprecipitated with acetone. After drying under reduced pressure, 7 gm of the polymer is isolated (35% yield).

The polymer is described as soluble in water and methanol but insoluble in ethanol and other common solvents. The product, if isolated by freeze drying, contains one molecule of water per repeat unit. On drying at 100°C and 0.1 mm Hg the water is lost but some anhydride formation takes place.

F. POLYMERIZATION OF SALTS OF THE ACRYLIC ACIDS

It is self-evident that salts of the poly(acrylic acids) may be produced either by the neutralization of the poly(acrylic acids) with bases, i.e., hydroxides, methylates, oxides, carbonates, bicarbonates, basic salts, and the like, or by the polymerization of preformed salts of the monomeric acids, either in solution or in the solid state. The neutralization of polymeric acids with polymeric bases such as poly(vinylpyridine) should also be considered.

Particularly in the case of neutralization of polymeric acids and of copolymers containing such acids, a wide range of products may be formed by partial neutralization of the carboxylate function.

The incorporation of methacrylic or acrylic acid in emulsion copolymers used in floor polishes has a long history. Among the properties leading to the use of coatings containing these components is their increased water solubility upon treatment with caustic. Thus the polish coating may be removed with soap and water.

The acrylates and methacrylates of divalent metals will form insoluble polymeric salts. Similarly, treatment of the polymeric acids with bases derived from divalent elements may result in crosslinked systems. Such crosslinks have been termed "temporary" crosslinks as distinguished from "permanent" ones, i.e., those containing covalent carbon–carbon bonds [16]. It may very well be that pigments such as titanium dioxide are bonded in an emulsion polymer system by such temporary crosslinks.

The reaction rates of all the acrylates decay rapidly at low conversions. The chain length of poly(potassium acrylate) is one order of magnitude greater than that of poly(lithium acrylate).

In the methacrylate series, sodium methacrylate reacts more rapidly than potassium methacrylate. Lithium methacrylate is entirely inert. Initially the polymerization of sodium methacrylate exhibits an accelerated reaction stage.

It is postulated that the difference in polymerization rates of the same unsaturated anions in the presence of different cations must be caused by the geometry of the crystal lattices involved. There is also evidence that the polymers lie in an amorphous phase while the reactive ends of the growing chains are anchored in the monomer lattice [17].

Calcium acrylate dissolves readily in water to give a reasonably stable solution at room temperature. At 24°C, 44 gm of calcium acrylate will dissolve in 100 gm of water. The freezing point of such a solution is approximately $-12°C$ [18].

The interesting phenomenon about this monomer, which finds application in soil stabilization, is the insolubility of the polymer in water although water dissolves in the polymer. Under equilibrium conditions, at room temperature, the hydrated polymer contains 25% of water. This polymer is rigid. Higher water concentrations may be achieved. The resultant swollen polymers, depending on the water content, may be mechanically hard, semirigid, soft, or very soft and rubberlike. The swollen polymers are not water-permeable.

NOTE: A recent article on "smart" or "intelligent" gels reported that certain hydrogels, containing materials such as poly(acrylic acid), poly(acrylamide), poly(vinylpyrrolidone), and polyethylene oxide–polypropylene oxide such as some of the Pluronics, react to external stimuli. Such systems, depending on their composition, of course, may respond to small temperature changes, changes in ionic strength, pressure, solvents, and electric or magnetic fields [19].

2. POLYMERIZATION OF ACRYLAMIDE AND RELATED AMIDES

Acrylamide, methacrylamide, N-methylolacrylamide (as a 60% aqueous solution), N,N'-methyleneacrylamide, and diacetoneacrylamide are amide-function monomers which have achieved a certain measure of commercial importance. Other related monomers have also been made available. However, by far the most important monomer of this group is acrylamide. Its polymer and copolymers find application in adhesives, dispersants, flocculants, viscosity modifiers and thickeners, additives for improved dying of fibers, protective colloids, and a host of other applications. Copolymers of acrylamide prepared in aqueous media usually are hydrogels. These have a variety of applications.

Interestingly enough, water solutions migrate readily into and out of such hydrogels. We have found that such hydrogels are readily loaded with aqueous solutions of ferrous and ferric salts.

Then, treatment with aqueous alkali precipitated paramagnetic compositions *in situ*. The resultant gel particles responded quite well to simple magnets.

Acrylamide and its simple analogues are generally soluble in water and also in a variety of nonaqueous solvents. They may also act as solvents for other materials including many inorganic salts. This property may be carried over to some extent to the polymers bearing amide functionalities.

Under certain polymerization conditions, imide formation may take place (Eqs. 1 and 2). Polymers which are partially imidized exhibit reduced water solubility. The intermolecular imide represents a thoroughly crosslinked polymer which is quite insoluble, while the intramolecular imides may exhibit only partial insolubility [20, 21].

$n(CH_2=CHCONH_2) \longrightarrow$

$$-CH_2-\underset{\underset{NH_2}{\overset{\overset{}{C=O}}{|}}}{CH}-CH_2-\underset{\underset{NH_2}{\overset{\overset{}{C=O}}{|}}}{CH}-CH_2-\underset{\underset{NH_2}{\overset{\overset{}{C=O}}{|}}}{CH}-CH_2-\underset{\underset{NH_2}{\overset{\overset{}{C=O}}{|}}}{CH}- \longrightarrow$$

Polyacrylamide

$$-CH_2-\underset{\underset{NH_2}{\overset{\overset{}{C=O}}{|}}}{CH}-CH_2-\underset{\overset{}{C=O}}{CH}\overset{CH_2}{\diagdown}\underset{\overset{}{C=O}}{CH}-CH_2-\underset{\underset{NH_2}{\overset{\overset{}{C=O}}{|}}}{CH}- +NH_3 \quad (1)$$

Intramolecular imidization

The molecular weight and extent of imidization may be controlled by adjustment of the concentration of the monomer in a solvent, changes in the pH of the solution, control of the reaction temperature, and use of water-soluble alcohols as chain-transfer agents [22].

$$-CH_2-\underset{\underset{NH_2}{\overset{\overset{}{C=O}}{|}}}{CH}-CH_2-\underset{\underset{NH_2}{\overset{\overset{}{C=O}}{|}}}{CH}-CH_2-\underset{\underset{NH_2}{\overset{\overset{}{C=O}}{|}}}{CH}-$$

$$+$$

$$-CH_2-\underset{\underset{NH_2}{\overset{\overset{}{C=O}}{|}}}{CH}-CH_2-\underset{\underset{NH_2}{\overset{\overset{}{C=O}}{|}}}{CH}-CH_2-\underset{\underset{NH_2}{\overset{\overset{}{C=O}}{|}}}{CH}- \longrightarrow$$

$$-CH_2-CH-CH_2-CH-CH_2-CH-$$
$$\underset{NH_2}{\overset{C=O}{|}} \quad \underset{NH}{\overset{C=O}{|}} \quad \underset{NH_2}{\overset{C=O}{|}}$$
$$\overset{C=O}{|} \quad +NH_3$$
$$-CH_2-CH-CH_2-CH-CH_2-CH-$$
$$\underset{NH_2}{\overset{C=O}{|}} \qquad \qquad \underset{NH_2}{\overset{C=O}{|}}$$

Intermolecular imidization (2)

With a propagation rate at 25°C of 1.8×10^4 liter/mole/sec, a high rate of polymer formation may be expected. The heat of polymerization of the aqueous monomer is approximately 19.5 kcal/mole [23]; consequently, adequate means of dissipating the evolving heat have to be found when more than a few grams of the monomer are to be polymerized. The heat of polymerization of methacrylamide is on the order of 13.5 kcal/mole.

The chemistry of acrylamide and related monomers has been reviewed [24–28]. Work on reactivity ratios based on three parameters—resonance stabilization energy, radical electronegativity, and monomer electronegativity—is reported by Hoyland [29].

The polymerization of acrylamide has been carried out by free-radical and anionic means using a variety of initiating systems. An unusual method of polymerization is the hydrogen-transfer process to form poly(β-alanine).

Safety considerations. Since acrylamide is supplied as a white crystalline material (m.p. 84°–85°C) its appearance is similar to many well-crystallized, pure, and innocuous organic products.

Only relatively recently has it been discovered that acrylamide, and possibly also the substituted acrylamides, are toxic in a unique manner. Neuromuscular disorders of varying severity have been reported. The monomer may be absorbed through the intact skin, by inhalation, or by ingestion. There may be a degree of cumulative toxicity in animals [26]. The 1989 Material Safety Data Sheet (MSDS) of the American Cyanamid Company states that acrylamide has caused cancer in laboratory animal tests [27]. The monomer is, therefore, considered a "cancer suspect agent." Reference [27] gives extensive details on exposure control methods. It may be wise to consider precautions in handling any of the acrylamide-related monomers. It is anticipated that during the 1990s the handling regulations will become much more stringent. Even relatively small bulk quantities of acrylamide may have to be manipulated in totally enclosed and properly engineered systems. It is suggested that the most current revisions of data sheets of Refs. [26, 27] be consulted when working with acrylamides and/or the related monomers.

Polyacrylamide, by the way, seems to be nontoxic provided no residual monomer is present.

As indicated above, the heat of copolymerization of the monomer and its rate of polymerization are high. Consequently heat-transfer problems exist in handling the monomer when even modest quantities of acrylamide are polymerized.

A. SUSPENSION POLYMERIZATION

Since acrylamide and methacrylamide are quite water-soluble solids, the usual technique of suspension polymerization which consists of heating monomer droplets, suitably initiated, in a stirred aqueous medium cannot be used to prepare homopolymers of these compounds. This, of course, does not mean that copolymers cannot be prepared by standard methods provided the distribution coefficient of the acrylamide in the comonomer and in water is favorable.

Alternative procedures involve adding a saturated aqueous solution of acrylamide converted to a water-in-oil emulsion by means of a suitable nonionic surfactant, an organic solvent, adding initiators and heating until beads of polymer form [30, 31]. By this technique, the beads probably are highly hydrated polyacrylamide.

Preparation 10-6 is an example of this technique taken from the patent literature. To be noted is the technique of removing the water in the course of the polymerization by azeotropic distillation. Probably one of any number of nonionic surfactants (e.g., Triton® X-405) may be substituted for the fatty acid amide used here.

This particular preparation had been developed before the current concerns with carcinogenicity and problematic environmental issues. Therefore, a substitute for benzene as an azeotroping agent and for carbon tetrachloride as a chain-transfer agent will have to be incorporated in this procedure.

10-6. Suspension Copolymerization of Acrylamide and Acrylic Acid [32]

$$n\text{CH}_2=\text{CH}-\overset{\overset{\text{O}}{\|}}{\text{C}}-\text{NH}_2 + m\text{CH}_2=\text{CH}-\overset{\overset{\text{O}}{\|}}{\text{C}}-\text{OH} \longrightarrow$$

$$\left(\begin{array}{c} -\text{CH}_2\text{CH}- \\ | \\ \text{C}=\text{O} \\ | \\ \text{NH}_2 \end{array} \right)_n \left(\begin{array}{c} -\text{CH}_2-\text{CH}- \\ | \\ \text{C}=\text{O} \\ | \\ \text{OH} \end{array} \right)_n \quad (3)$$

In an apparatus fitted for azeotropic removal of water and with a stirrer, to a refluxing mixture of 225 gm of benzene, 1 gm of carbon tetrachloride, and 14 gm of ethoxylated tallow fatty acid amides is added over a period of approximately 2 hr a solution of 91.8 gm of acrylamide, 39.2 gm of acrylic acid, 16 gm of 25% aqueous solution of sodium vinylsulfonate, 10 gm of urea, 30 gm of water, 37 gm of 25% aqueous ammonia, and 0.02 gm of potassium persulfate. Heating is continued until all of the water has been removed. The bead polymer is filtered off and dried under reduced pressure at 50°C. Yield: 134 gm of a water-soluble copolymer.

B. SOLUTION POLYMERIZATION

From the standpoint of laboratory procedures, solution polymerizations of acrylamide and of related monomers are probably most satisfactory. Aqueous reaction media are widely used, generally with those free-radical initiators which are water-soluble.

Polymerizations at low pH tend to lead to imidization, while very high pH conditions may lead to hydrolysis of the amidic functional groups. A variety of additives have been used with the basic initiator systems. Among these are ammonia, ammonium chloride, and ammonia-producing compounds [33]. These reagents are said to reduce the induction period of the polymerization, increase the rate of polymerization, and produce polymers of higher molecular weight than usual which are still readily dispersed in water. They are stable to degradation in dilute aqueous solution and yield polymers with superior flocculating properties when compared with polymers prepared in the absence of ammonia. Alcohols may be added to the initial reaction charge to act as chain-transfer agents [26]. Other chain-transfer agents, such as mercaptans, may also be used. Salts of the alkali metals or of the alkaline earth metals as well as 3,3′,3″-nitrilotris(propionamide) enhance the rate of polymerization and control the chain length [34]. Preparation 10-7 is an example of the general procedure used for aqueous solution polymerizations. The method was taken from a patent.

The use of 3,3′,3″-nitrilotris(propionamide), triethanol amine, dimethylpropionitrile, etc., seems to have the effect of activating the initiation process in such a manner that the product develops molecular weights of 12 million or more, whereas ordinary solution polymers yield polyacrylamide of a molecular weight of 6 million [35].

10-7. Polymerization of Acrylamide in Aqueous Solution [34]

In a 1 liter flask, to a solution of 27 gm of acrylamide and 54 gm of sodium sulfate in 210 ml of deoxygenated water under nitrogen is added with stirring 0.012 gm of 3,3′,3″-nitrilotris(propionamide) and 0.012 gm of ammonium persulfate (see Note). The reaction mixture is warmed at 60°C for 30 min. The solution is then cooled. In the product, 94% of the monomer has been converted to polymer.

NOTE: Many aqueous solution polymerizations are carried out with persulfate salts. Whereas ammonium persulfate is frequently recommended, this particular initiator deteriorates rapidly on storage at room temperature. Consequently we suggest that potassium persulfate or sodium persulfate be substituted. Potassium persulfate is a salt with relatively low solubility in water, whereas sodium persulfate is quite soluble.

10-8. Preparation of Poly(acrylamide) Gels

In the preparation of poly(acrylamide) gels for electrophoresis, the activator of choice is another tertiary amine, N,N,N',N'-tetramethylethylene diamine (TEMED) [36].

In poly(acrylamide) gel electrophoresis (PAGE) for the separation of nucleic acids, proteins, and other charged species, the crosslinking agents are added to a solution of acrylamide in an aqueous buffer. By far the most common crosslinking agent used is N,N'-methylenebisacrylamide (BIS or MBA). The resulting gel slab is insoluble ("irreversible gel").

For applications where the separated bands of protein are to be recovered, crosslinking agents that have been used are ethylene diacrylate, N,N'-diallyl tartardiamide, N,N-(1,2-dihydroxyethane)-bis-acrylamide, and N,N',N''-triallyl citrictriamide, all of which may be hydrolyzed to break up the crosslinks that hold the gel together. Bis-acrylyl cystamine (N,N'-cystamine-bis-acrylamide) is a crosslinking agent that may subsequently be cleaved with a sulfide [36]. A typical gel recipe to prepare one 0.75 mm × 10 cm × 14 cm slab or twelve 7.5 cm column tubes is:

11.0 ml	of distilled water;
2.0 ml	of a Tris-Borate–EDTA buffer (pH 8.3);
4.0 ml	of a solution; 0.76 gm of acrylamide and 0.04 gm of MBA in water (i.e., a 20% acrylamide: MBA solution in which the ratio of acrylamide to MBA is 19 to 1);
2.0 ml	of a 1% TEMED solution in water.

These solutions are mixed at room temperature in the order given and degassed for 0.5 min. Then polymerization is initiated, after placing the mixture in the appropriate apparatus, by adding 1 ml of a freshly prepared 2% solution of ammonium persulfate followed by gentle swirling. The gel is allowed to polymerize overnight [36].

In organic solvents, the polymerization of acrylamide may be carried out with the usual organic initiators. In many cases the polymer precipitates from the reacting system. Often the resultant products are water insoluble.

A 20% solution of acrylamide in dioxane has been polymerized at 60°C in the presence of 0.3% of benzoyl peroxide. By nitrogen analysis of the polymer some imide formation was detected [37].

A procedure for the preparation of polymethacrylamide in toluene is given here as an example of a technique which is said to produce a water-soluble polymer.

10-9. Polymerization of Methacrylamide in Toluene [38]

NOTE: The following procedure is carried out near the boiling point of toluene and the melting point of methacrylamide

(110°–111°C). Since the initiator benzoyl peroxide is quite unstable at this temperature, it is introduced into the reaction system as a dilute room temperature solution in toluene. Due safety precautions must be taken.

Behind a safety shield, a 200 ml flask is fitted with a nitrogen inlet tube, a reflux condenser, a mechanical stirrer, and a buret for the addition of the initiator solution. Under a nitrogen atmosphere, a solution of 15.0 gm of methacrylamide in 45 ml of toluene is placed in the flask. The flask is then mounted in an oil bath that is thermostatted at 120°C ± 1.5°C. Over a 3 hr period, while a slow stream of nitrogen passes over the solution (care being taken that the gas stream does not cause excessive losses of the boiling solvent), a solution of 0.36 gm of benzoyl peroxide in 13 ml of toluene is added in 10 equal portions. After the additive has been completed, heating is continued for another 0.5 hr. The polymer is then filtered off and heated three times with three fresh portions of 30 ml of methylene dichloride. The polymer is dried to constant weight under reduced pressure with occasional warming to 100°C. It is finally dried over calcium chloride and paraffin under reduced pressure. Yield: 11.8 gm (78%), intrinsic viscosity $[\eta]$ = 0.60 dl/gm (c = 0.764).

C. EMULSION POLYMERIZATIONS

(a) Oil-in-Water Emulsions. Since acrylamide is water soluble and polyacrylamide is also water soluble, true polymer lattices in water cannot form during homopolymerization processes. Certain copolymers have, however, been prepared in emulsion or microemulsion form.

Copolymerizations of styrene with acrylamide in surfactant-free aqueous media are of particular interest because of their potential application to the immobilization of proteins. Of additional interest is the fact that the latex particles generated are quite uniform in particle size.

The amido groups attached to the polymer chains may be subjected to a variety of typical organic reactions to enhance reactivity of the polymeric particles to proteins [39–44].

Procedure 10-10 outlines the general method of copolymerization used by Kawaguchi and co-workers [39–44]. In this series of reports it is stated that the mechanism involves three stages: First, the acrylamide polymerizes preferentially in the aqueous phase; then, as the conversion of acrylamide levels off, the styrene polymerizes (or copolymerizes) in particles; and finally, the residual acrylamide polymerizes. At a total conversion of 80% of the combined monomers, somewhat less than 40% of the acrylamide present in the original monomer composition will be found in the final polymer. The papers in question are not clear as to whether the water-soluble resin had been separated from the precipitated polymer.

10-10. Emulsion Copolymerization of Styrene and Acrylamide [39, 40]

In a 300 ml flask fitted with a stirrer, a gas inlet tube, and reflux condenser, 12 kg of styrene, 8 gm of acrylamide, and 150 ml of deionized water, buffered to a pH of 9, are subjected to a slow stream of nitrogen for 1 hr. Then 10 ml of a potassium persulfate (5 mmole per liter of water) is added. The mixture is heated, with stirring, at 70°C for 3 hr. The latex formed has particles of 0.31 micrometers diameter, conversion ca. 80%. The polymer may be isolated by pouring the latex into an excess of acetone, filtering the polymer off, and drying the product under the reduced pressure. We estimate that the resulting polymer contains ca. 20% acrylamide (i.e., ca. half of the charged acrylamide is not in the copolymer but probably was lost to the water layer).

(b) Water-in-oil emulsions (inverse polymerization). The inverse emulsion polymerization of acrylamide has been extensively discussed in Ref. [45]. Interestingly, in their procedure, 2,2'-azobisisobutyronitrile (AIBN) is used as the initiator. This initiator is soluble in the organic phase but limited in solubility in water. Evidently the latex generation depends on very small amounts of AIBN-based free radicals occasionally entering aqueous acrylamide droplets. Thus a process somewhat analogous to ordinary latex polymerizations takes place. Procedure 10-11 is an example of an inverse emulsion polymerization. To be noted is that a mixture of low HLB and a high HLB surfactant is used here, with the concentration of the low HLB surfactant being much larger than that of the high HLB surfactant. More recent procedures (cf. Ref. [46]) actually use 50-50 mixtures of two such surfactants. In Procedure

10-11, surfactants manufactured by Seppic Society are used. The Montane 83 is a sorbitan sesquioleate similar to Span 60. The Montanox 85 is a high-HLB emulsifier, polyoxyethylene sorbitan tri-oleate, analogous to Tween 85.

10-11. Inverse Emulsion Polymerization of Aqueous Acrylamide in Toluene [45]

A 1.5 liter round-bottomed reactor fitted with a motor-driven steel paddle-type stirrer, nitrogen inlet tube, temperature control, and reflux condenser is placed in a water bath which may be thermostatted so that the reaction system is maintained at 50°C.

To this reactor is charged 300 gm of toluene, 34 gm of Montane 83, and 1.4 gm of Montanox-80. Then a solution of 40 gm of acrylamide in 60 gm of deionized water is added. The mixture is stirred for 30 min at 1,000 rpm while being purged with nitrogen. The emulsion is equilibrated at 50°C, and the stirrer is slowed to 250 rpm. The AIBN, 20 mg, is dissolved in a minimum amount of toluene. The AIBN solution is then added to the emulsion and the process allowed to proceed for approximately 6 hr. A stable emulsion forms. Conversion of monomer to polymer is estimated to be 86%, the weight average molecular weight to be 2.56 million.

To isolate the polymer, the emulsion may be diluted with additional toluene. The dilute solution is added to an excess of acetone to precipitate the polymer. The polymer is filtered off and dried under reduced pressure.

Microemulsions are spontaneously formed dispersions of water and oil phases in the presence of appropriate emulsifiers. They are said to be transparent and thermodynamically or kinetically stable. The droplet size ranges from 10 to 100 Å in diameter [46, 47]. An inverse microemulsion polymerization recipe has been given by Schulz [46].

A 250 gm aqueous solution containing 106 gm of acrylamide, 15.9 gm of sodium acetate, and 300 gm of "Isoparfin" fraction [b.p. 207–254°C] is treated with a mixture of 22 gm of Sorbitan sesquioleate and 135 gm of Sorbitan hexoleate (HLB of mixture = 9.3). To the composition is added 0.32 gm of AIBN. The mixture is cooled to 19°C and subjected to ultraviolet radiation. After 15 min, a stable transparent inverse microlatex forms. In the case of such inverse microemulsions, the rate of polymerization is said to be 10–200 times faster than inverse emulsion polymerization because of the great number of very small particles generated by the high level of surfactant [46].

D. ANIONIC POLYMERIZATION

The polymerization of acrylamide in the presence of simple anionic reagents such as sodium *tert*-butylate in the presence of free-radical inhibitors leads primarily to poly(β-alanine) by hydrogen-transfer polymerization, a topic to be discussed in the next section.

N-Substituted and N,N-disubstituted acrylamides in some cases may be polymerized by reagents such as metal alkyl–transition element halide catalysts or alkyl metals to produce crystalline polymers. Among the first crystalline materials produced by this means was poly(N-isopropylacrylamide) [48], followed by a series of crystalline N,N-disubstituted acrylamides [49].

$$\text{Na} + \text{CH}_2{=}\text{CH}\overset{\overset{\displaystyle O}{\|}}{\text{C}}{-}\text{NR}_2 \longrightarrow \text{Na}^+ + \cdot\text{CH}_2{-}\overset{\displaystyle |}{\underset{\underset{\displaystyle NR_2}{|}}{\underset{\displaystyle C=O}{|}}}{\overline{\text{CH}}}{:} \xrightarrow{\text{monomer}}$$

$$\text{Radical} \underset{\text{atactic block}}{\sim\!\sim\!\sim\!\sim} \text{CH}_2{-}\underset{\displaystyle CONR_2}{\overset{\displaystyle |}{\text{CH}}} \underset{\text{isotactic block}}{\sim\!\sim\!\sim\!\sim} \text{anion} \quad (4)$$

The polymers have a certain degree of tacticity, although the polymer chains are probably block copolymers of stereospecific polymer segments accompanied by free-radical chain segments. The polymerization process may be pictured as in Eq. (4) [49].

The catalysts suitable for anionic polymerization are based on the metals of the first and second groups of the periodic table in hydrocarbon media. In general, the lower the dielectric constant of the solvent, the greater the degree of tacticity of the polymer.

10-12. Anionic Polymerization of N,N-Dimethylacrylamide [49]

In a flask fitted with a stirrer and an argon inlet is placed a solution of N,N-dimethylacrylamide in toluene. The flask is thermostatted at room temperature, the air is displaced with argon, and a solution of ethyllithium is added. The temperature in the flask rises rapidly and within a few seconds the swollen polymer precipitates. The reaction is terminated by the addition of methanol. The product is filtered off and dried under reduced pressure. The yield is said to be quantitative. The product is said to be tactic and highly crystalline.

E. HYDROGEN- (OR PROTON-) TRANSFER POLYMERIZATION

In a patent, Matlack [50] disclosed the formation of poly(β-alanine) when acrylamide was heated in the presence of bases. With subsequent investigations a novel method of polymerization was developed which has received much attention.

The incentive for extensive research in this field, aside from that generated by academic interest, may be attributed to the fact that poly(β-alanine), may

$$\left[CH_2-CH_2-\overset{O}{\overset{\|}{C}}-NH \right]_n$$

be considered Nylon-3. Indeed it can be converted to fibers. The frequency of amide groups in the polymer chain is great, and the polymer is expected to resemble natural silk more closely than other nylons. However, it would appear that the molecular weight of these polymers is usually too low to permit commercial development.

The reaction of acrylamide to give poly(β-alanine) may be considered an example of the Michael reaction.

The initiation of the reaction may proceed by one of the two reactions represented by Eqs. (5) and (6) [51]. Equation (5) is the one favored by most investigators.

$$CH_2=CHCONH_2 + B^- \rightleftharpoons CH_2=CHCONH + BH \qquad (5)$$

or

$$CH_2=CHCONH_2 + B^- \rightleftharpoons B-CH_2CHCONH_2 + BCH_2CH_2CONH^- \qquad (6)$$

The propagation step also may be visualized as having two possible courses, the intramolecular propagation (Eq. 7) or the intermolecular propagation (Eq. 8):

$$CH_2=CHCONH^- + CH_2=CHCONH_2 \longrightarrow$$

$$CH_2=CH-CONHCH_2CHCONH_2 \longrightarrow CH_2CH_2CONHCH_2CH_2CONH \quad (7)$$

Intramolecular propagation

$$CH_2=CHCONH + CH_2=CHCONH_2 \longrightarrow$$

$$CH_2=CHCONHCH_2-CHCONH_2 + CH_2=CHCONH_2 \longrightarrow$$

$$CH_2=CHCONHCH_2CH_2CONH_2 + CH_2=CH\overline{CONH} \quad (8)$$

Intermolecular propagation

Intermolecular propagation has been considered the major factor in explaining the experimental details which have been accumulated [52, 53].

In their extensive review of the hydrogen-transfer polymerization, Kennedy and Otsu [52] state that besides the step growth by the proton-transfer mechanism, anionic chain growth also takes place. The products therefore are either mixtures or copolymers formed by the two possible mechanisms. The difficulties in reaching clear-cut decisions as to the details of the process may arise from the fact that the polymerizations are carried out at temperatures above 100°C, so that it is difficult to differentiate between the various processes. Molecular weights and conversions are usually fairly low even upon prolonged heating.

Preparation 10-13 is an example of proton-transfer polymerization. In such preparations, it is normal to inhibit the possibilities of free-radical polymerization by incorporation of phenyl-β-naphthylamine (PNA) (**CAUTION:** May be carcinogenic). However, since PNA is derived from naphthylamine, which is no longer available, another inhibitor will have to be developed.

10-13. Hydrogen-Transfer Polymerization of Acrylamide in Pyridine [51]

To a solution of 100 ml of specially dried pyridine containing 0.02 gm of phenyl-β-naphthylamine (**CAUTION:** May be carcinogenic), heated to 100°C, is added, with mechanical stirring, 10 gm of sublimed acrylamide. After the monomer has been dissolved, a solution of 0.1 gm of sodium in 10 ml of *tert*-butyl alcohol is added.

Within a few minutes some polymer begins to separate. Heating and stirring is continued for 16 hr.

The polymer is then filtered off and extracted with boiling water for 1 hr. The insoluble portion is dried under reduced pressure at 80°C. Yield: 4.8 gm (48%).

The aqueous extract is neutralized with acetic acid and evaporated to dryness. The residue is dissolved in water and precipitated by stirring into methanol (yield 2.6 gm). From the mother liquor 0.4 gm of acrylamide and 1.2 gm of another solid may be isolated.

3. POLYMERIZATION OF ACRYLONITRILE AND RELATED MONOMERS

Polymers and copolymers of acrylonitrile are of considerable economic importance. Poly(acrylonitrile) and certain copolymers may be drawn into fibers (e.g., "Orlon," acrylic fiber, modacrylic fiber, etc.) which compete with wool in textile applications. However, in 1990, Du Pont, the second largest producer of Orlon in the United States, with a capacity of 130 million kg of this fiber annually, ended acrylic fiber production [54]. At the present time, Orlon-type textiles are difficult to find on the consumer market.

The use of butadiene-acrylonitrile copolymers (Buna rubber, Perbunan, etc.) as rubber substitutes goes back to pre–World War II days. Because of their resistance to various solvents, these nitrile rubbers have found wide application on their own merit. Copolymers of acrylonitrile and styrene (SAN resins) as well as terpolymers of acrylonitrile, butadiene, and styrene and blends containing ABS resins are useful molding compounds. Elastomers of acrylonitrile, styrene, and acrylic monomers are known as ASA elastomers. Lattices of copolymers containing acrylonitrile have a wide variety of applications.

An acrylonitrile polymer has been used as the starting material for the preparation of one variety of graphite fibers. These are important in the production of high-performance reinforced plastics, particularly in the aerospace industry.

Methacrylonitrile has become of some importance since a commercially useful method of manufacture has been developed. Its behavior in polymerization is somewhat analogous to that of acrylonitrile. The major characteristic which differentiates the polymerization processes of methacrylonitrile from those of acrylonitrile is that poly(acrylonitrile) is virtually insoluble in monomeric acrylonitrile, while poly(methacrylonitrile), like other acrylic polymers in relation to their monomers, is soluble in methacrylonitrile. The softening point of poly(methacrylonitrile) is low, as compared to that of poly(acrylonitrile). Consequently, poly(methacrylonitrile) is not suitable for the production of fibers.

Since solvent and oil resistance of the nitrile polymers is associated with the nitrile function, a monomer with a higher concentration of this grouping per molecule is naturally desirable. For this reason, vinylidene cyanide has been studied from time to time. However, this molecule has such violent physiological effects that work with it should be carried out with extreme precautions (see Sandler and Karo [55]). The monomer polymerizes readily in the presence of moisture. The residual monomer in a sample of poly(vinylidene cyanide), even after approximately 2 weeks in a normal laboratory atmosphere, was sufficiently potent to cause a relapse of asthma-like symptoms, which one of our associates had contracted upon initial exposure to a trace of the monomer [56].

The esters of α-cyanoacrylic acid polymerize under basic conditions. In particular, methyl α-cyanoacrylate is commercially produced as a metal-to-metal adhesive ("Superglue"). Its action depends on a polymerization initiated by a moisture layer on the metal surfaces.

The literature on nitrile monomers is voluminous. The producers of the monomers have supplied useful reviews on acrylonitrile [57] and on methacrylonitrile [58, 59]. An extensive review will also be found in the *Encyclopedia of Polymer Science and Technology* [60].

Safety and hazard considerations. Acrylonitrile is a toxic monomer and is designated a carcinogen [61]. We consider it prudent that all nitrile monomers be handled as if they were highly toxic and possibly carcinogenic. Ingestion, inhalation, and absorption of liquids and vapors through the skin and mucous membranes must be prevented. Reference [61] gives details about the toxicology and about emergency first aid procedures, personal protection, shipping and handling, storage, waste disposal, and the handling of spills and leaks for acrylonitrile. This information should also be applied to the handling of related nitrile monomers unless specific information for any given compound reduces the handling requirements. In any case, the current manufacturers' Material Safety Data Sheet (MSDS) for each compound must be reviewed prior to use.

A. BULK POLYMERIZATION

The bulk polymerization procedures for these monomers are similar to those described for acrylate esters. However, in the case of acrylonitrile, only relatively small quantities of the monomer can be polymerized in bulk.

Table II lists typical physical properties of both acrylonitrile and methacrylonitrile. To be noted are the relatively high heats of polymerization, the composition of the monomer/water azeotropes, and the solubilities of the monomers in water as well as the solubility of water in the monomers.

The factors which differentiate the free-radical polymerization characteristics of acrylonitrile from those of many other monomer systems are the inhibition by oxygen, the rather high heat of polymerization, the fact that the polymer is insoluble in its monomer, the high solubility of the monomer in water, and the extensive solubility of water in the monomer. In regard to most of these particular factors, methacrylonitrile behaves more like typical acrylic monomers.

Since the polyacrylonitrile is insoluble in its monomer, at even low conversions heavy slurries form from which heat transfer is particularly difficult. As a result, soon after the polymerization process has been initiated, auto-acceleration may take place, and the process may become uncontrollable.

Typical initiators which have been used on a small scale to initiate acrylonitrile polymerizations have included benzoyl peroxide [62], 2,2′-azobis(isobutyronitrile) and 2,2′azobis(2,4-dimethylvaleronitrile) [63], tetraalkyldiarylethanes [64], silver salts dissolved in monomer [65], and hydrogen peroxide [66]. Polymerizations have also been carried out in the vapor phase [67].

B. SUSPENSION POLYMERIZATION

Since acrylonitrile has an appreciable solubility in water, suspension homopolymerizations of this monomer, at best, may be expected to behave as a mixture of solution and suspension processes. This may be described as a slurry polymerization.

TABLE II
PHYSICAL PROPERTIES OF ACRYLONITRILE AND METHACRYLONITRILE

Property	Acrylonitrile[a]	Methacrylonitrile[b]
Molecular weight	53.06	67.09
Freezing point (°C)	−83.55 ± 0.05	−35.8
Refractive index, n_D^{25}	1.3888	1.3977
Surface tension (dynes/cm) (24°C)	27.3	24.4
Boiling point (°C/760 mm)	77.3°	90.3°
(°C/50 mm)	8.7°	—
Density (gm/ml at 25°C)	0.8004	d_4^{25} 0.805
Azeotropes		
Benzene	b.p. 73.3°C, 47% by wt. acrylonitrile	—
Isopropanol	b.p. 71.7°C, 56% by wt. acrylonitrile	—
Water	b.p. 71°C, 88% by wt. acrylonitrile	b.p. 77°C, 84% by wt. methacrylonitrile
Entropy of polymerization kcal/mole		−34 ± 2
Activation energy of polymerization (kcal/mole)	—	5.4
Heat of polymerization (bulk) (kcal/mole)	−17.3 ± 0.5	−13.5 ± 0.2 (−15.3 ± 1)
Softening temperature of polymer	Decomposes above 200°C	100°−130°C Depolymerizes at 250°C (amorphous polymer)

	% H_2O in acrylonitrile	% Acrylonitrile in water	% H_2O in methacrylonitrile	% Methacrylonitrile in water
Solubility in water				
at 0°C	2.1	7.2	1.06	2.89
20°C	3.1	7.35	1.62	2.57
40°C	4.8	7.9	2.38	2.59
50°C	6.3	8.3	2.83	2.69
80°C	10.8	11.0		

[a] From Ref. [57].
[b] From Ref. [58].

10-14. Slurry Homopolymerization of Acrylonitrile [68]

In a four-necked 2 liter resin kettle fitted with a nitrogen inlet tube, addition funnel, stirrer, thermometer, and condenser is placed 900 gm of freshly boiled and cooled deionized water and 0.029 gm of concentrated sulfuric acid. The stirrer is started and a rapid stream of nitrogen is passed through the solution for 20 min while the reactor is maintained with a constant temperature bath at 35°C.

Then 53.0 gm of acrylonitrile is added while maintaining a rapid nitrogen stream for an additional 10 min. The nitrogen flow is reduced and the gas inlet tube is raised to just above the surface of the stirred reaction mixture. Five hundred milliliters of an aqueous solution containing 1.71 gm of sodium persulfate (the original reference called for an equivalent amount of ammonium persulfate) is added slowly followed by the slow addition of a solution of 0.71 gm of sodium metabisulfite in 50 ml of water.

After about 3 min, the solution becomes cloudy. Some heat is evolved during the first 30 min. Gradually a thick slurry develops. After 4 hr the stirring is stopped and the polymer is filtered off, washed, in turn, with 1 liter of deionized water, and then with 175 ml of methanol. The polymer is dried overnight at 70°C under reduced pressure. Yield: 48 gm (92%).

C. SOLUTION POLYMERIZATION

Since acrylonitrile has appreciable water solubility, the boundaries between suspension, solution, and emulsion polymerization of this monomer are ill defined. We have already indicated under the heading of suspension and slurry polymerization that it is quite common to start with an aqueous solution of acrylonitrile, and, because of the insolubility of the polymer, to produce a thick slurry.

Solvents for polyacrylonitrile are of considerable importance in the production of fibers and films from polyacrylonitrile. Table III lists a variety of materials in which polyacrylonitrile is soluble [69].

An even more extensive list of solvents as well as nonsolvents for polyacrylonitrile and poly(methacrylonitrile) will be found in the *Polymer Handbook* [70].

Quite generally, since monomeric acrylonitrile is soluble in a wide variety of organic solvents, if polymerization of the monomer is carried out in one of the common solvents, the polymer precipitates soon after it forms. If the solution involves a good solvent for the polymer, naturally polymer solutions are formed. By way of contrast, the solubility behavior of methacrylonitrile and of polymethacrylonitrile is quite different. For example, a toluene solution of this polymer may be prepared by heating a solution of methacryl-

TABLE III
Solvents for Polyacrylonitrile[a]

66.7% Lithium bromide in water (at 100°C)	γ-Butyrolactone
	β-Hydroxypropionitrile
Concentrated aqueous solutions of sodium thiocyanate or zinc chloride	1,3,3,5-Tetracyanopentane
	Dimethyl sulfoxide
75% Sulfuric acid	Tetramethylene sulfoxide
Dimethylformamide	2-Hydroxyethyl methyl sulfone
Dimethyl sulfone	Methyl ethyl sulfone
Sulfolane	m-Nitrophenol
Dimethylacetamide	p-Nitrophenol
Dimethylthioformamide	Phenylene diamine (o-, p-, or m-)
N-Methyl-β-cyanoethylformamide	Methylene dithiocyanate
α-Cyanoacetamide	Tetramethylene dithiocyanate
Tetramethyloxamide	Dimethyl cyanamide
Malononitrile	Hot α-chloro-β-hydroxypropionitrile
Fumaronitrile	Hot N,N-di(cyanomethyl)aminoacetonitrile
Succinonitrile	Hot ethylene carbonate
Adiponitrile	Hot propiolactone
ε-Caprolactam	Hot succinic anhydride
Bis(β-cyanoethyl) ether	Hot maleic anhydride
Hydroxyacetronitrile	

[a] From Ref. [69].

onitrile in toluene with AIBN at 70°C [71] and a solution polymer of methacrylonitrile in tertiary butanol has been prepared by straightforward solution polymerization with benzoyl peroxide at 79°C [72]. With solvents such as DMF or DMSO, inorganic initiators such as potassium persulfate or persulfate-bisulfite may be used as initiators for the polymerization of acrylonitrile.

10-15. Solution Polymerization of Acrylonitrile in Zinc Chloride Solution [73]

In a 250 ml flask, 72 gm of zinc chloride is dissolved in 18 gm of water. After displacing the air with nitrogen, and at 80°C, a solution of 0.002 gm of benzoyl peroxide in 10 gm of acrylonitrile is added with stirring. Heating and stirring at 80°C under nitrogen are continued until a highly viscous solution is produced. This solution may be extruded into water at 70°–80°C to produce a fiber whose tensile strength may be increased by stretching.

D. EMULSION POLYMERIZATION

The homopolymerization of acrylonitrile by the emulsion technique is somewhat difficult, while copolymers, particularly those containing low levels of acrylonitrile, and methacrylonitrile polymers and copolymers form stable latexes more readily.

The difficulty of emulsion polymerizations of acrylonitrile may be attributed both to the insolubility of the polymer and to the water solubility of the monomer. In the emulsion polymerization procedure, part of the acrylonitrile may polymerize in aqueous solution while the remainder may tend to form a true latex. Obviously this situation will tend to form polymers which are not homogeneous as to molecular weight.

10-16. Emulsion Polymerization of Acrylonitrile [74]

In a 5 liter, three-necked flask fitted with a nitrogen inlet tube, stirrer, thermometer, and condenser, supported in a constant temperature bath at 35°C, is placed 3100 ml of distilled water, 40 gm of sodium lauryl sulfate (weight on dry solids basis), and 20 ml of 0.1 N sulfuric acid. Nitrogen is bubbled through this solution rapidly for 15 min. Then the flow of nitrogen is reduced to approximately 3 bubbles/sec and 800 gm of acrylonitrile is added with stirring. After 10 min, a solution of 4.0 gm of fresh ammonium persulfate in 50 ml of water and a solution of 1.82 gm of sodium metabisulfite in 50 ml of water is added.

During the first hour, the temperature in the flask is maintained at 35°C by adding ice to the water bath. Thereafter the bath needs to be heated. Within 3 hr the polymerization is substantially complete. The latex is filtered through a cheese cloth.

Two recipes that produced monodisperse polyacrylonitrile lattices consisted of:

1 liter	deionized water,
0.270 gm	potassium persulfate, and
40.0 gm	acrylonitrile.

The composition was reacted, with stirring, for 20 min at 50°C. The resulting latex particles had a mean diameter of 100 nm with standard deviation of 10 nm.

In a recipe consisting of

1 liter	deionized water,
0.135 gm	potassium persulfate,
0.138 gm	sodium lauryl sulfate,
30 gm	acrylonitrile, and
0.013 gm	sodium metabisulfite,

the composition was maintained for 24 hr at 50°C. The resulting latex had mean particle diameters of 226 nm with a standard deviation of 8 nm.

The lattices were dialyzed for 20 days with daily changes of deionized water, presumably to remove water-soluble components of the aqueous phase [75]. In our own experience with other polymer lattices, this slow dialysis procedure is of dubious efficiency, particularly as far as removing surfactant is concerned. We have found treatment of lattices with an ion-exchange resin is more efficient. Recently, we have exchanged the aqueous serum of a latex by use of a Millipore "Pellicon" tangential flow filter system.

These were used as lattices for the seeded heterogeneous polymerization of acrylonitrile using either potassium persulfate or γ-irradiation to initiate the process. It was found that the polymerization took place on the seed particle surface, the average number of free radicals per particle was high, and the free-radical capture efficiency by the particles was 100% [75].

Characteristically, the emulsion polymerization of methacrylonitrile requires a longer period of time. The process can be carried out with a higher concentration of monomer in water than in the case of acrylonitrile.

10-17. Emulsion Polymerization of Methacrylonitrile [76]

In a sturdy 500 ml bottle is placed 225 gm of distilled water, 3.0 gm of sodium lauryl sulfate, 1.0 gm of n-dodecyl mercaptan, 100 gm of methacrylonitrile, and 0.5 gm of potassium persulfate. Nitrogen is bubbled through the mixture, the bottle is closed, placed in a protecting perforated metal sleeve, and shaken at 55°–70°C for

6-10 hr. The bottle is cooled to room temperature and cautiously opened. The latex is filtered through cheese cloth. Yield: greater than 90%.

E. ANIONIC AND STEREOSPECIFIC POLYMERIZATION

Both acrylonitrile and methacrylonitrile may be polymerized in the presence of anionic initiators. Since the propagation step is susceptible to termination by protons, these reactions are normally carried out in aprotic solvents. Acrylonitrile does not give rise to useful products readily. Its polymers tend to be yellow, unless produced at very low temperatures, and, under the alkaline reaction conditions, side reactions involving cyanoethylation may take place.

A large variety of initiators for the anionic polymerization have been studied. Among these are alfin catalysts [77], alkoxides [78], butyllithium [79], metal ketyls [80], solutions of alkali metals [81], sodium malonic esters [82], sodium hydrogen sulfide and potassium cyanide [83], sodium in liquid ammonia or sodium amide [84], sodium cyanide [85], and calcium oxide [86].

10-18. Monosodium Benzophenone-Initiated Polymerization of Acrylonitrile [80]

(a) Preparation of monosodium benzophenone. In a three-necked flask equipped with a high-speed stirrer, provisions for introducing highly purified nitrogen, a condenser protected against atmospheric moisture and oxygen, and a thermometer, to 50 ml of highly purified tetrahydrofuran is added 1.48 gm of sodium wire and 1.15 gm of benzophenone with rapid stirring in a nitrogen atmosphere. A blue color appears almost instantaneously. Rapid stirring is continued for 15 min. There is then said to be no residual sodium.

(b) Polymerization of acrylonitrile. To the flask containing the monosodium benzophenone solution, enough purified THF is added to obtain 60 ml of solution. The stirred solution is cooled to 0°C and 10 ml (8 gm) of acrylonitrile is added at once. The temperature in the flask is maintained at 0°C for 1 hr. Then a mixture of 10 ml of concentrated hydrochloric acid and 10 ml of methanol is cautiously added. The precipitated polymer is filtered off and washed in turn with dilute hydrochloric acid, water, methanol, and ether. It is then dried under reduced pressure at 60°C.

4. POLYMERIZATION OF ACRYLIC AND METHACRYLIC ESTERS

The polymers and copolymers of acrylic and methacrylic acids have found a wide range of applications that is continuously expanding. Among these uses are methacrylate glazing, protective coatings, paint bases, binders, adhesives, elastomers, floor polishes, leather finishes, plastic bottles, and dental filling materials, to mention only a few.

The methods of polymerization of these esters—by bulk, in solution, in suspension, and as emulsions—are also applicable to many of the other common "vinyl" monomers, such as styrene, vinyl acetate, vinyl chloride, vinylidene chloride, acrylonitrile, acrylamide, and homologues, analogues, and derivatives of these classes of compounds.

The suspension and emulsion procedures are probably the best methods of polymerizing reasonable quantities of material since they resemble standard procedures used in the organic preparative laboratory.

A. GENERAL CONSIDERATIONS

Since a large number of acrylic esters are available, polymers with a range of properties are readily prepared. Naturally, by resorting to co- and terpolymerization techniques, the properties of the resulting polymers may be modified further so that the time-worn phrase "tailor-making" resins for specific applications does indeed have considerable validity. The more recently studied "interpenetrating polymers" further broaden the range of properties and applications of vinyl polymers.

A few guidelines for developing products with specific properties may be briefly summarized thus:

1. Acrylate esters, with the exception of methyl acrylate, generally produce elastomeric polymers, whereas methacrylates of corresponding alcohols are usually more brittle or more resinous in nature. As a guide, the glass-transition temperature, T_g, is the physical constant which must be considered when flexibility and other elastomeric properties are of importance. Roughly speaking, materials with a T_g below 0°C are elastomeric [87].

2. The average molecular weight, the molecular weight distribution, and the degree of chain branching affect properties. Thus, for example, one may anticipate differences in injection molding properties between a resin with a normal Gaussian distribution of the molecular weight of the polymer chains and one with the same average molecular weight, but consisting of a blend with two or more resins of different individual average molecular weights which have been compounded to give the same composite average molecular weight.

3. The chemical nature of the monomers affects properties of their polymers in a great variety of ways. For example, the chain length of the alcohol portion of an acrylic ester has an effect on the glass-transition temperature and the brittle point. Similarly, the percent elongation generally increases with increasing chain length while the tensile strength decreases. The percent elongation of polyacrylates is substantially greater than that of polymethacrylates as a class [88]. This may be related to the "legginess" of polyacrylate-based adhesive solutions. On the other hand, the tensile strengths of polymethacrylates are generally higher than those of the corresponding polyacrylates [88].

While acrylic polymers exhibit good oxidation resistance, differences may be expected between acrylates and methacrylates. Variation in crosslinking properties may be related to the spacing between two acrylic units attached to a glycolic chain and/or to the number of polymerizable units in the monomer (e.g., compare the effect of ethylene dimethacrylate with that of the tetramethacrylate of pentaerythritol). Allyl methacrylate is another interesting crosslinking agent since the two polymerizable moieties react at different rates with a substrate methacrylate. This permits crosslinking of a methacrylate as a second stage of a molding process.

Monomers such as 2-hydroxyethyl methacrylate, 2-N,N-dimethylaminoethyl methacrylate, acrylic acid, methacrylic acid, and many other monomers will affect the properties of a copolymer such as sensitivity to hydration or to changes in pH [89]. The presence of halogens in a monomer may affect the refractive index of its polymer as well as its flame resistance.

4. Chemical modification of a polymer system may influence properties. For example, polymers derived from α-hydroxyethyl

acrylic esters may be crosslinked with a variety of polyfunctional reagents which react with alcohols such as diisocyanates. Glycidyl acrylic ester polymers may react under conditions similar to those used with epoxy resins.

5. While an extensive discussion of co- and terpolymerization is beyond the scope of this chapter, it should be mentioned here that properties of polymers may be modified by the judicious copolymerization of several monomers. The sequence of monomers influences the properties of the polymer. Control of this arrangement of monomer units depends, in part, on the copolymerization ratio of pairs of monomers. By special techniques, "block" and "graft" copolymers may be produced.

6. Stereospecific polymers may be produced under certain conditions. Such products may exhibit significantly different properties from their amorphous atactic counterparts.

7. Blends of different polymers may be produced.

8. A host of reinforcing fillers, plasticizers, surfactants, solvents, modifiers, foaming agents, etc., may be compounded with a parent polymer to modify and control desired properties.

9. The method of polymerization also is of importance in tailor-making a product. For example, if the end product is to be used as a coating material, an emulsion polymerization technique will be selected. On the other hand, poly(methyl methacrylate) glazing may have to be prepared by bulk polymerization in a suitable mold.

The polymerization techniques will affect the molecular weight, the molecular weight distribution, and the degree of branching of the polymer. In emulsion polymerizations, particle-size distribution, latex stability, etc., are controlled by the method of polymerization.

B. REACTANTS AND REACTION CONDITIONS

(a) Inhibitors. As normally supplied, acrylic esters are inhibited to enhance their storage stability. The most common inhibitors used with these monomers are hydroquinone (HQ) and *p*-metho-

xyphenol (MEHQ). The concentration level of these inhibitors is usually in the range of 50–100 parts per million by weight.

(b) Initiators. In the polymerization of acrylic monomers, diacyl peroxides and azo compounds are most commonly used in bulk, suspension, and organic solution processes. Combinations of oxidizing and reducing agents such as hydroperoxides and the formaldehyde-bisulfite addition product, or persulfates, either by themselves or in combination with bisulfites or a host of other reducing agents, find application in emulsion and aqueous solution polymerization.

Probably the most widely used initiator for the polymerization of acrylic monomers is dibenzoyl peroxide (BPO). For reasons of safety in transportation, this compound is usually supplied as a dispersion containing 30–40% of water.

The selection of an initiator is to a large extent an empirical matter. With the exception of emulsion polymerization procedures, acrylic ester polymerizations normally require initiators which are soluble in the monomer. Particularly in suspension polymerizations, the initiator must have a very low solubility in water to reduce the possibility of latex formation. On the other hand, in emulsion polymerizations classically the initiating system had to be low in monomer solubility but high in water solubility (e.g., the sodium metabisulfite–sodium persulfate redox system). Quite satisfactory redox initiation systems have been devised for emulsion processes in which monomer-soluble hydroperoxides are incorporated in the monomer while a water-soluble reducing agent is also used. Hydrogen peroxide, which has a surprisingly high solubility in certain monomers, has also been used in conjunction with water-soluble reducing agents to produce latexes.

2,2′-Azobis(isobutyronitrile) (AIBN) and other azo compounds can be used at relatively low temperatures (60°–80°C). AIBN has the added advantage that it is not an oxidizing agent and therefore is useful in the production of colorless polymers. Since its thermal decomposition is relatively simple, polymers formed from its free radicals are thought to be free of branched chains.

C. CHAIN-TRANSFER AGENTS

A variety of compounds may act to reduce the average molecular weight of the polymer by a chain-transfer mechanism during polymerization. Solvents may act as chain-transfer agents, although their activity is usually low. The most commonly used agents are mercaptans, particularly the higher molecular weights ones such as dodecyl mercaptan.

Halogenated compounds such as carbon tetrachloride and chloroform have particularly high chain-transfer constants. These compounds must be used with extreme caution since explosive polymerizations have been observed.

D. REACTION TEMPERATURE

The reaction temperature of a polymerization has a profound influence on the course of the reaction. Other factors being equal, the higher the reaction temperature, the lower the average molecular weight of the polymer produced.

However, the temperature also has an effect on the rate of decomposition of initiators, the number of effective free radicals produced, and the reactivity of these free radicals. Temperature also affects chain-transfer agent activity. These considerations may offset the simple reciprocal relationship between temperature and molecular weight.

The viscosity of the reacting system, of course, is also temperature dependent. The diffusion of monomers and of growing polymer chains as well as the heat-transfer properties of the system are modified as the viscosity of the reacting medium increases or as the molecular weight changes.

E. POLYMERIZATION PROCEDURES

Most of the common acrylic esters may be homopolymerized by relatively simple procedures. Variations in the methods that are used may be made because of requirements related to the final applications of the product, limitations set by available equipment, the reactivity of the monomers, and the physical state and other properties of the monomer and of the polymer.

(a) Bulk polymerization. The conversion of a monomer to a polymer in the absence of diluents or dispersing agents is termed a "bulk" polymerization.

From the laboratory standpoint, this procedure may be the simplest method since it basically consists of heating a solution of an initiator in the monomer.

In the casting of polymer sheets, bulk polymerization of initiator-containing monomers may be used, although "prepolymers" (i.e., solutions of polymers in their monomers, usually prepared by bulk polymerizing a monomer until the viscosity of the mixture has reached a desired level) [90] are more commonly used commercially to reduce problems arising from shrinkage, not to mention breakage of monomer through the flexible gasketing usually used. The method of making polymer sheets has been reviewed in some detail by Beattie [91]. While his review shows the method of producing large castings, the technique does lend itself to use in the laboratory.

10-19. Generalized Procedure for the Preparation of Polymer Sheets

A casting mold is constructed by clamping a length of polyethylene tubing between three sides of two sheets of carefully cleaned plate glass with spring paper clips of suitable size. It is most convenient to make the mold of plate glasses cut to the same width but unequal lengths (e.g., one piece 10×10 cm and one 10×15 cm) and forming the gasket along three sides. The excess glass section will facilitate the filling of the mold.

The mold is supported in an inclined position with the larger side forming the lower part of the mold. A solution of methyl methacrylate containing 0.5% of benzoyl peroxide is carefully poured into the mold to fill approximately two-thirds of the available space.

The mold is then supported in an upright position, preferably in a shallow dish. The top of the mold may be closed by forcing a length of tubing along the open edge. The assembly is then placed in an *explosion-proof*, high-velocity air oven and heated at 70°C for 72 hr. The curing time varies with the overall size of the casting. In the case of thick-cross-section castings, the curing time is considerably longer.

After the polymerization has been completed, the casting is cooled gradually. The sheet is removed by removing the glass plates. Because of the inhibiting effect of air, the top portion of the sheet may be somewhat soft. The plastic sheet may be finished by conventional plastic shaping techniques.

Similar molding techniques may be used with casting syrups, although great care must be exercised to allow trapped bubbles to rise prior to the final curing stages. With sizable glass molds, special arrangements have to be made to prevent bulging of the glass plates under the hydrostatic pressure of the monomer.

(b) Suspension polymerization. A sharp distinction must be drawn between suspension (or slurry) and emulsion polymerization processes.

We use the term *suspension polymerization* to refer to the polymerization of macroscopic droplets in an aqueous medium to produce macroscopic polymer particles. These are readily separated from the reaction medium. The polymerization kinetics are essentially those of a bulk polymerization, with the expected adjustments associated with carrying out a number of bulk polymerizations in small particles more or less simultaneously and in reasonably good contact with a heat exchanger (i.e., the reaction medium) to control the exothermic nature of polymerization. Usually the process is characterized by the use of monomer-soluble initiators and of suspending agents.

On the other hand, emulsion polymerization processes involve the formation of colloidal polymer particles which are essentially permanently suspended in the reaction medium. The reaction mechanism involves the migration of monomer molecules from liquid monomer droplets to sites of polymerization which originate in micelles consisting of surface-active agent molecules surrounding monomer molecules. For present purposes it is enough to state here that emulsion polymerizations are usually characterized by the requirement of surfactants during the initiation of the process and the use of water-soluble initiators. This process also permits good control of the exothermic nature of the polymerization.

An estimate of the approximate temperature rise during a polymerization may be made as follows: The heats of polymerization of many monomers are published data. For acrylic monomers, the figure is approximately 20 kcal per mol. A calorie has been defined as the quantity of heat that will raise the temperature of one gram of water, at room temperature, one degree Celsius. Thus 20 kcal will raise the temperature of 1 kg of water by 20°C.

Suspension polymerizations are among the most convenient laboratory as well as plant procedures for the preparation of polymers. The advantages of the method include wide applicability (it may be used with most water-insoluble or partially water-soluble monomers), rapid reaction, ease of temperature control, ease of preparing copolymers, ease of handling the final product, and control of particle size.

In this procedure, the polymer is normally isolated as fine spheres. The particle size is determined by the reaction temperature, the ratio of monomer to water, the rate and efficiency of agitation, the nature of the suspending agent, the suspending agent concentration, and, of course, the nature of the monomer.

Common suspending agents are polyvinyl alcohols of various molecular weights and degrees of hydrolysis, starches, gelatin, calcium phosphate (especially freshly precipitated calcium phosphate dispersed in the water to be used in the preparation), salts of poly(acrylic acid), gum arabic, gum tragacanth, etc. In some cases surfactants have been used to suspend the monomer (especially in the case of fluorinated acrylates which usually are difficult to polymerize in emulsion processes).

The initiators usually used are dibenzoyl peroxide, lauryl peroxide, 2,2'-azobis(isobutyronitrile) and others which are suitable for use in the temperature range of approximately 60°-90°C.

The suspension process may be carried out not only with compositions consisting of a solution of the initiator in the monomer, but also with complex mixtures which incorporate plasticizers, pigment particles, chain-transfer agents, modifiers, etc.

10-20. Suspension Polymerization of Methyl Methacrylate [92, 93]

In a laboratory hood, to a 500 ml three-necked flask, equipped with a sealed stirrer and *explosion-proof* motor, condenser, thermometer, and an addition funnel is placed 150 ml of a 1% solution of sodium poly(methacrylate) in water and 5 ml of an aqueous solution containing 0.85 gm of disodium phosphate and 0.05 gm of monosodium phosphate. A solution of 0.5 gm of dibenzoyl peroxide in 50 gm of inhibitor-free methyl methacrylate is placed in the flask. Into the addition funnel is measured 25 ml of a 5% aqueous solution of sodium poly(methacrylate). The reaction flask is placed in a water bath and stirring is started. The stirring rate is adjusted to such a speed that monomer droplets form that have a diameter between 2 and 3 mm. The reaction flask is heated at approximately 80°C for

60 min. The temperature is controlled by raising or lowering the bath (**CAUTION**). If a tendency to agglomeration is observed, small amounts of the suspending agent in the addition funnel are added.

After the polymerization has gone to completion, the product is collected on a funnel fitted with cheese cloth. The polymer is washed with fresh boiling water until freed of water-soluble materials. The product is dried under reduced pressure at 60°–70°C.

(c) Solution polymerization. The polymerization of acrylic monomers in solution is usually carried out quite simply [88]. It must be kept in mind that since the viscosity of a polymer solution increases with increasing molecular weight, limitations are imposed on this method by the handling problems involving high-viscosity systems. Many solvents for the monomer also may act as chain-transfer agents. Consequently the development of very high molecular weights in a solution process is difficult.

In organic solvents, "oil-soluble" initiators, such as dibenzoyl peroxide, are used. In aqueous solutions, sodium persulfate has been suggested.

10-21. Solution Copolymerization of Glycidyl Methacrylate and Styrene [94]

To 228 gm of xylene maintained at 136°C is added with stirring, over a 3 hr period, a solution of 453 gm of styrene, 80 gm of glycidyl methacrylate, and 11 gm of di-*tert*-butyl peroxide. After the addition has been completed, the solution is heated and stirred for an additional 3 hr at 136°C (to 100% conversion). After cooling, the polymer solution may be diluted to 54% solids by the addition of 228 gm of methyl isobutyl ketone. The relative viscosity of the copolymer (1% solution in 1,2-dichloroethylene at 25°C) is 1.175.

The photopolymerization of isopropyl acrylate in solution at −105°C has been mentioned [95]. The procedure leads to a syndiotactic polymer. In a hexane solution at 40°C syndiotactic poly(methyl methacrylate) is said to form by polymerization with a sodium dispersion, while *n*-butyllithium or phenylmagnesium bromide at −40°C in toluene leads to isotactic poly(methyl methacrylate) [96–99].

Polymerization of methyl methacrylate by radical initiation at 25°C in the presence of preformed isotactic poly(methyl methacrylate) in dimethylformamide solution has resulted in the formation of syndiotactic polymers that had formed a complex with the isotactic resins [100].

Monomers such as 2-hydroxyethyl methacrylate are distinctly soluble in water. Their polymerization in aqueous systems is of considerable importance. From the industrial standpoint, perhaps the most valuable application arises from the fact that the polymer, when slightly crosslinked, may be prepared in such a manner that it traps the water in which the reaction is carried out to produce a soft elastic gel. Such systems find application in the preparation of soft contact lenses and as carriers for the localized application of medications. The example given here to illustrate the method of preparation has been patented and is for reference only.

10-22. Preparation of a Crosslinked Polymer Gel [101]

To a dispersion of 150 gm of acrylamide, 100 gm of 2-hydroxyethyl methacrylate, 0.1 gm of ethylene dimethacrylate, and 750 gm of water is added 10 ml of a 2% aqueous solution of sodium thiosulfate and 15 ml of a 2% aqueous solution of ammonium persulfate.

The mixture is poured into an appropriate mold and allowed to polymerize at room temperature. The product is described as a soft, elastic material which may be washed in running water for several hours to remove water-soluble materials. To preserve the elastic nature of the material, it should be stored in contact with water.

NOTE: See earlier comments on the hazards of handling acrylamide.

(d) Emulsion polymerization. In the section on suspension polymerization we have indicated the differentiation between suspension and emulsion (or latex) polymerization. Emulsion polymers usually are formed with the initiator in the aqueous phase, in the presence of surfactants, and with polymer particles of colloidal dimensions, i.e., on the order of 0.1 μm in diameter [102]. Generally the molecular weights of the polymers produced by an emulsion process are substantially greater than those produced by bulk or suspension polymerizations. The rate of polymer production is also higher. Since a large quantity of water is usually present, temperature control is often simple.

It is beyond the scope of this work to discuss theoretical aspects of emulsion polymerization. References [102–116] contain material pertinent to the theoretical and practical aspects of this topic.

While most of these references deal primarily with the Harkins–Smith–Ewart approach to the interpretation of the emulsion polymerization process, alternative mechanisms have been proposed [102].

In the preparation of a polymer latex, the initial relationship of water, surfactant, and monomer concentration determines the number of particles present in the reaction vessel. Once the process is underway, further addition of monomer does not change the number of latex particles.

A more detailed discussion of the factors involved in the preparation of emulsion polymers (lattices) will be found in a separate chapter.

From the preparative standpoint there are two classes of initiating systems:

1. The thermal initiator system, in which use is made of water-soluble materials which produce free radicals at a certain temperature to initiate polymerization. The most commonly used materials for such thermal emulsion polymerizations are potassium persulfate, ammonium persulfate, or sodium persulfate.

2. The activated or redox initiation systems. Since these systems depend on the generation of free radicals by the oxidation–reduction reactions of water-soluble compounds, initiation near room temperature is possible. In fact, redox systems operating below room temperature are available (some consisting of organic hydroperoxides dispersed in the monomer and a water-soluble reducing agent). A typical redox system consists of ammonium persulfate and sodium metabisulfite. There is some evidence, particularly in the case of redox polymerizations, that traces of iron salts catalyze the generation of free radicals. Frequently these iron salts are supplied by impurities in the surfactant (quite common in the case of surfactants specifically manufactured for emulsion polymerization) or by stainless steel stirrers used in the apparatus. In other recipes, the iron salts may be supplied in the form of ferrous ammonium sulfate, or, if the pH is low enough, in the form of ferric salts.

It has been recommended that a small amount of 70% t-butyl hydroperoxide be added immediately after the reducing agent has been added (approx. 3.5 to 5 drops per 200 gm of monomer, followed by another few drops of t-butyl hydroperoxide at the peak of the reaction temperature) [110].

While emulsion polymers have been prepared by shaking all components in a sealed ampoule [117] or by stirring all components together [118] the slightly more complex procedure given in Procedure 10-23 is usually more satisfactory.

10-23. Emulsion Polymerization of Ethyl Acrylate (Thermal Initiation) [88, 110]

In a 2 liter Erlenmeyer flask, in a hood, to 800 ml of deionized water is added in sequence 96 gm of Triton X-200, 1.6 gm of ammonium persulfate, and 800 gm of ethyl acrylate. The contents is thoroughly mixed to form a monomer emulsion.

In a 3 liter resin kettle fitted with a stainless steel stirrer, a thermometer which extends well into the lowest portion of the reactor, a reflux condenser, and a dropping funnel a mixture of 200 ml of deionized water and 200 ml of the monomer emulsion is heated in a water bath while stirring at a constant rate in the range of 160–300 rpm to an internal temperature of 80°–85°C. At this temperature refluxing begins and vigorous polymerization starts (frequently signaled by the appearance of a sky-blue color at the outer edges of the liquid). The temperature may rise to 90°C. Once refluxing subsides, the remainder of the monomer emulsion is added from the dropping funnel over a 1 to 2 hr period at such a rate as to maintain an internal temperature between 88° and 95°C by means of the hot water bath, if necessary. After the addition has been completed and the reaction temperature has subsided, the temperature of the latex is raised briefly to the boiling point. Then the reaction mixture is cooled to room temperature with stirring. After removing stopcock grease from all joints, the latex is strained through a fine-mesh nylon chiffon. Only a negligible amount of coagulum should be present.

In particular, if a latex is to be used for coatings, adhesives, or film applications, no silicone-based stopcock greases should be used on emulsion polymerization equipment. While hydrocarbon greases are not completely satisfactory either, there are very few alternatives.

In the above examples, it will be noted that the monomer is added to the reaction system as an oil-in-water emulsion. Many emulsion polymerizations are more simply carried out by adding pure monomer to an aqueous dispersion of surfactant and initiators. This procedure permits a more rigid control of the number of particles in the aqueous phase.

The redox initiation system may consist of a large variety of reagents. Usually it consists of a water-soluble oxidizing agent such as ammonium persulfate and a reducing agent such as sodium metabisulfite, sodium hydrosulfite, or sodium formaldehyde sulfoxylate, and an activator, usually a salt of a polyvalent metal, particularly ferrous sulfate or ferrous ammonium sulfate. As we have indicated before, traces of metallic ions such as one would obtain from metal stirrers may be sufficient to act as activators. While water-soluble oxidizing agents are usually used, monomer-soluble hydroperoxides have also been used in redox systems along with a variety of reducing agents including reducing sugars and metal activators. The book by Bovey *et al.* [103], while it reviews the World War II experience with the polymerization of butadiene–styrene elastomers, is still a useful introduction to this area of redox polymerization.

10-24. Redox Emulsion Polymerization of Ethyl Acrylates [110]

In a 1 liter resin kettle fitted as in Procedure 10-23 except that a nitrogen inlet tube is also attached to the apparatus is prepared a solution of 376 ml of deionized water and 24 gm of Triton X-200. While the nitrogen flow is on, with stirring, 200 gm of uninhibited ethyl acrylate, 4 ml of a solution freshly prepared by dissolving 0.3 gm of ferrous sulfate heptahydrate in 200 ml of deionized water, and 1 gm of ammonium persulfate are added. After stirring for 30 min, the mixture is cooled to 20°C and 1 gm of sodium metabisulfite and 5 drops of 70% *tert*-butyl hydroperoxide are added. The polymerization starts rapidly and the temperature in the flask rises to nearly 90°C. After approximately 15 min, the polymerization is complete. The latex is cooled and passed through a nylon chiffon strainer.

If inhibited monomers are used, an induction period is observed in this process although the rate of polymerization remains about the same. Variations of procedure such as the gradual addition of monomer to a seed polymer are also possible. The gradual addition of reducing agent solution along with gradually added monomer may also be used as a technique of redox latex polymerization.

(e) *Ionic polymerization.* In recent years, anionic polymerizations of acrylic monomers have received considerable attention, although commercial application of stereospecific polymers of acrylic monomers produced by such procedures seem to be lacking [95, 96, 98, 99, 118–125].

Syndiotactic polymers of methyl methacrylate are produced in highly solvating media with 9-fluorenyllithium while the isotactic polymers are formed by organolithium compounds in hydrocarbon solvents or by Grignard reagents in toluene. The syndiotactic–isotactic stereoblocks form in moderately solvating media (ether or toluene containing ether or dioxane) with 9-fluorenyllithium, dibutylmagnesium, or diphenylmagnesium in toluene at 0° or at −70°C [126]. It would seem that these preparative systems are such that the products formed are really mixtures or complexes of several of the stereoregular types in which one or the other predominates and lends itself to more or less ready separation.

10-25. Preparation of Isotactic Poly(methyl methacrylate) [127]

In a 500 ml four-necked flask with suitable adapters to carry a mechanical stirrer, addition funnel, thermometer, inlets and outlets for dry, oxygen-free nitrogen, and an addition port covered with a rubber serum-bottle cap, is placed 300 ml of anhydrous toluene. Inhibitor-free, dry methyl methacrylate (15 gm, 0.15 mole) is placed in the addition funnel and nitrogen is bubbled through both the monomer and the solvent for 3 hr. After cooling the flask contents to 3°C with an ice bath, 3.6 ml of a 3.3 M solution of phenylmagnesium bromide in diethyl ether is injected by means of a hypodermic syringe through the serum cap, while, with moderate stirring, the monomer is added. Stirring is continued for 4 hr. The viscous solution is then poured slowly into 3 liters of vigorously stirred petroleum ether. The solid is collected on a filter and dried under reduced pressure. Inorganic impurities are removed by digestion with aqueous methanol containing hydrochloric acid. Yield, 11.2 gm (74.6%) of isotactic poly(methyl methacrylate), viscosity average molecular weight = 480×10^3. After annealing at 115°C, crystalline characteristics of the polymer can be demonstrated by X-ray diffraction.

Over a toluene solution of methyl methacrylate, with proper precautions about the exclusion of moisture and oxygen, was layered a solution of butyllithium as a separate phase in a sealable glass tube. After sealing, the layered system was allowed to remain at a predetermined polymerization temperature (e.g., $-74°C$) for 24 hr. The system was then cooled in liquid nitrogen and the tube was carefully broken. The resin was cut into three approximately equal sections. Analysis of the polymer showed that by this slow polymerization technique approximately 80% of the polymer was isotactic [128]. On the other hand, under more ordinary reaction conditions, again with the exclusion of moisture and oxygen, 10 mmol of methyl methacrylate in 10 ml of toluene was polymerized in the presence of 0.6 mmol of triethylaluminum and 0.2 millimol of t-butyllithium at $-75°C$ for 24 hr. The resulting polymer was said to be 92.3% syndiotactic. Without the trialkylaluminum compound, the polymer was only 5.8% syndiotactic [128].

In a study of the polymerization of perdeuterated ethyl methacrylate in toluene with butyllithium at $-76°C$, the polymer was found to consist of syndiotactic, isotactic, and oligomeric polymers. The polymer mixture was extracted with methanol. The insoluble fraction was the syndiotactic PEMA. The methanol soluble fraction was treated with methanol/water (97/3). The soluble portion of the extraction was treated with hexane. The insoluble fraction from this treatment was identified as the isotactic polymer, and the soluble fraction was oligomeric [129]. It is interesting to note that the separation of an isotactic from syndiotactic polymer could be accomplished simply on the basis of differences in their solubility in simple solvents.

For simple demonstration experiments of anionic polymerizations where precision of kinetic or structural data is not required, much less elaborate techniques of purification and handling may be satisfactory. We have found that after careful flaming of a simple apparatus under high-purity nitrogen, careful drying of the monomers with calcium hydride, transferring the monomers to the apparatus with the usual serum cap and syringe procedure, and redrying the monomers with a trace of *sec*-butyllithium solution before injecting the main portion of the initiator solution, reasonable anionic polymerizations can frequently be carried out [130].

(f) Group transfer polymerization. The "group transfer" reaction of silyl ketene acetals as applied to unsaturated conjugated carbonyl compounds such as acrylic esters becomes a "living" polymerization, a process that is a new approach to the design and construction of polymer chains [131–133]. Equations (9) and (10) outline the process as visualized by Webster and co-workers [133]. For clarity we have marked the important carbon atoms of the methyl methacrylate involved in the first step with an asterisk (*). Note also the resemblance of the circled unit to structural features of methyl methacrylate.

$$CH_2=\overset{*}{C}-\overset{*}{\underset{CH_3}{\underset{*}{C}}}-OMe\ +\ \underset{CH_3}{\overset{CH_3}{C}}=\underset{OSiMe_3}{\overset{OMe}{C}}$$

$$\downarrow HF_2^-$$

$$\underset{*CH_3}{\underset{|}{CH_3-\overset{MeO}{\underset{|}{C^*}}\!\!\overset{\diagup O}{=}}} - CH_2 - \underset{CH_3}{\overset{MeO}{\underset{|}{C}}\!\!\overset{\diagup OSiMe_3}{}} \qquad (9)$$

$$(10)$$

$$\downarrow\ n\ CH_2=\underset{CH_3}{\overset{O}{\underset{|}{C}-\overset{\|}{C}-OMe}}$$

$$\underset{CH_3}{\underset{|}{CH_2-\overset{MeO}{\underset{|}{C}}\!\!\overset{\diagup O}{=}}}\!\!\left(\!\!-CH_2-\underset{\overset{|}{O}\!\!\overset{\diagup}{=}\overset{C}{\underset{\diagdown\!OMe}{}}}{\overset{CH_3}{\underset{|}{C}}}\!\!\right)_{\!n}\!\!\underset{CH_2}{\overset{CH_3}{C}}=\underset{OSiMe_3}{\overset{OMe}{C}}$$

This process has been termed *group transfer polymerization* since the trimethylsilyl group is transferred to the incoming monomer.

Procedure 10-26 is a brief description of a typical group transfer polymerization (GTP).

10-26. Group Transfer Polymerization of Methyl Methacrylate [133]

In a thoroughly dried apparatus, to a solution of 8.5 mmol of 1-methoxy-1-(trimethylsiloxy)-2-methyl-1-propene and 0.1 mmol of tris-(dimethylamino)sulfonium bifluoride in 20 ml of tetrahydrofuran at room temperature is slowly injected 360 mmol of methyl methacrylate with the exclusion of moisture. After the evolution of heat has subsided, 3 ml of methanol is injected. The product solution is removed from the apparatus and allowed to evaporate. A quantitative yield of PMMA is isolated. The number average molecular weight is 4,300, weight average molecular weight is 5,300 (theory, 4,343).

By the sequential addition of other monomers, before the process is "killed" with ethanol, block copolymers may be prepared. Reference [133] tabulates a number of di- and triblock polymers that have been prepared by this method. Monomers such as allyl methacrylate, glycidyl methacrylate, methyl vinyl ketone, and N,N-dimethylmethacrylamide have been polymerized by GTP procedures.

REFERENCES

1. V. A. Kargin, V. A. Kabanov, S. Ya. Mirlana, and A. V. Vlasov, *Vysokomol. Soedin* **3**, 134 (1961); *Polym. Sci. USSR* (*Engl. Transl.*) **3**, 28 (1962).
2. "Glacial Methacrylic Acid, Glacial Acrylic Acid," Bulletin CM-41 I/ej, Rohm and Haas Company, Philadelphia, Pennsylvania, Industrial Chemical Department, 1992.
3. A. G. Evans and E. Tyrrall, *J. Polym. Sci.* **2**, 387 (1947); P. Monjol and G. Champetier, *Bull. Soc. Chim. Fr.* No. 4, p. 1302 (1972).
4. G. Greber and G. Egle, *Makromol. Chem.* **40**, 1 (1960).
5. J. D. Borbey, D. J. T. Hill, A. P. Land, and J. H. O'Donnell, *Macromolecules* **24**, 2280 (1991).
6. S. R. Sandler and W. Karo, "Polymer Syntheses," 2nd ed., Vol. 1, pp. 336FF, Academic Press, San Diego, 1992.
7. S. C. Khanna, T. Jecklin, and P. Speiser, *J. Pharm. Sci.* **59**, 614 (1970).
8. J. C. Leyte and M. Mandel, *J. Polym. Sci., Part A* **2**, 1879 (1964).
9. S. Fakirov, D. Simov, R. Baldjieva, and M. Michailov, *Makromol. Chem.* **138**, 27 (1970).
10. Z. Priel and A. Silberberg, *J. Polym. Sci., Polym. Phys. Ed.* **8**, 689, 705, and 713 (1970).
11. C. J. Mast and W. R. Cabaness, *J. Polym. Sci., Polym. Lett. Ed.* **11**, 161 (1973).
12. N. Higashi, T. Nojima, and M. Niwa, *Macromolecules* **24**, 6549 (1991).
13. A. Katchalsky (Katzir) and G. Blauer, *Trans. Faraday Soc.* **47**, 1360 (1951).
14. S. Nagai and K. Yoshida, *Kobunshi Kagaku* **17**, 748 (1960); *Chem. Abstr.* **55**, 24086 (1961).
15. S. Nagai and F. Fujiwara, *Polymer Lett.* **7**, 177 (1969).
16. R. N. Bashaw and B. G. Harper, U.S. Patent 3,090,736 (1963).

17. H. Morawetz and I. D. Rubin, *J. Polym. Sci.* **57**, 669 (1962).
18. "Calcium Acrylate," Bulletin SP-42, Rohm and Haas Company, Special Products Department, Philadelphia, Pennsylvania, 1955.
19. See a review article with references by R. Dagani, *Chem. Eng. News* **75**(25), 26 (June 9, 1997).
20. W. O. Kenyon and L. M. Minsk, U.S. Patent 2,486,190 (1949); *Chem. Abstr.* **44**, 1750 (1950).
21. L. M. Minsk and W. O. Kenyon, U.S. Patent 2,486,192 (1949); *Chem. Abstr.* **44**, 1750 (1950).
22. L. M. Minsk, W. O. Kenyon, and J. H. Van Campen, U.S. Patent 2,486,191 (1949); *Chem. Abstr.* **44**, 1750 (1950).
23. R. M. Joshi, *J. Polym. Sci.* **60**, 556 (1962).
24. C. E. Schildknecht, "Vinyl and Related Polymers," p. 134ff, Wiley, New York, 1952.
25. W. M. Thomas, *Encycl. Polym. Sci. Technol.* **1**, 177 (1964).
26. "Chemistry of Acrylamide," Process Chemicals Department, American Cyanamid Co., Wayne, New Jersey, 1969.
27. M. A. Friedman, National Safety Data, MSDS No. 1528-08 (CAS No. 000079-06-1; Date: 05/23/89), American Cyanamid Co., Wayne, New Jersey.
28. "Methacrylamide," SP-32, Rohm & Haas Company, Speciality Products Department, Philadelphia, Pennsylvania, 1954.
29. J. R. Hoyland, *J. Polym. Sci., Part A-1* **8**, 885 and 901 (1970).
30. Morningstar-Paisley, Inc., British Patent 991,416 (1965).
31. Nalco Chemical Co., British Patent 1,000,307 (1965); J. W. Zimmermann and A. Kühlkamp, U.S. Patent 3,211,708 (1965).
32. H. Spoor, German Patent 2,064,101 (1972); *Chem. Abstr.* **77**, 127438u (1972).
33. M. B. Goren, U.S. Patent 3,200,098 (1965).
34. American Cyanamid Co., Netherlands Patent Appl. 6,505,750 (1965).
35. E. R. Kolodny, U.S. Patent 3,002,960 (1961).
36. Polysciences, Inc. Data Sheet #155 "Acrylamide, Electro/pure: Triply Recrystallized, Electrophoresis Grade," Warrington, PA, (June 1985) and references cited therein.
37. H. C. Haas and R. L. MacDonald, *J. Polym. Sci., Part A-1* **9**, 3583 (1971).
38. C. L. Arcus, *J. Chem. Soc., London* p. 2732 (1949).
39. Y. Ohtsuka, H. Kawaguchi, and Y. Sugi, *J. Appl. Polymer Sci.* **26**, 1637 (1981).
40. H. Kawaguchi, Y. Sugi, and Y. Ohtsuka, in "Emulsion Polymers and Emulsion Polymerization" (D. R. Bassett and A. C. Homielec, eds.), *ACS Symposium Series* **165**, 146 (1981).
41. H. Kawaguchi, Y. Sugi, and Y. Ohtsuka, *J. Appl. Polym. Sci.* **26**, 1649 (1981).
42. H. Kawaguchi, Y. Sugi, and Y. Ohtsuka, *J. Appl. Polym. Sci.* **26**, 2015 (1981).
43. H. Kawaguchi, H. Hoshino, H. Amagasa, and Y. Ohtsuka, *J. Colloid Interface Sci.* **77** (2), 465 (1984).
44. H. Kawaguchi, K. Sakamoto, Y. Ohtsuka, T. Ohtake, H. Sekguchi, and H. Iri, *Biomaterials* **10**, 225 (May 1989).
45. C. Graillat, C. Pichot, A. Guyot, and M. S. El Aasser, *J. Polym. Sci., Part A: Polym. Chem.* **24**, 427 (1986).

46. D. N. Schulz, Personal Communication to W. K. and *Am. Chem. Soc., Polymer Preprints* **30** (2), 329 (Sept. 1989).
47. M. Rosoff and A. Giniger, *Colloid and Interface Science* **V**, 475 (1976); M. Rosoff, *Surface and Membrane Science* **12**, 405 (1978); S. S. Atik and J. K. Thomas, *M. Am. Chem. Soc.* **104**, 5868 (1982); idem, *J. Am. Chem. Soc.* **105**, 4515 (1983).
48. D. J. Shields and H. W. Coover, Jr., *J. Polym. Sci.* **39**, 532 (1959).
49. K. Butler, P. R. Thomas, and G. J. Tyler, *J. Polym. Sci.* **48**, 357 (1960).
50. A. S. Matlack, U.S. Patent 2,672,480 (1954).
51. D. S. Breslow, G. E. Hulse, and A. S. Matlack, *J. Amer. Chem. Soc.* **79**, 3760 (1957).
52. J. P. Kennedy and T. Otsu, *J. Macromol. Sci., Part C* **6**, 237 (1972).
53. K. Yamaguchi and Y. Minoura, *J. Polym. Sci., Part A-1* **10**, 1217 (1972).
54. Anonymous, *Chem. Eng. News* **68**, 5 (June 18, 1990).
55. S. R. Sandler and W. Karo, "Organic Functional Group Preparations," 2nd ed., Vol. 1, Chapter 17, p. 571, Academic Press, New York, 1983.
56. B. D. Halpern, W. Karo, L. Laskin, P. Levine, and J. Zomlefer, Tech. Rep. WADC TR 54-264. Wright Air Development Center, Air Research and Development Command, United States Air Force, Wright-Patterson Air Force Base, Ohio, 1954.
57. "The Chemistry of Acrylonitrile," 2nd ed., Petrochemicals Department, American Cyanamid Company, New York, 1959.
58. "Methacrylonitrile," Technical Bulletin Vistron Corporation, Cleveland, Ohio, 1968; "Handling, Storage, Analysis of Acrylonitrile," American Cyanamid Company, Wayne, New Jersey, 1970.
59. "Methacrylonitrile," Patent Summary and Bibliography. Vistron Corporation, Cleveland, Ohio, 1967.
60. C. H. Bamford, G. C. Eastward, and A. Levovits, *Encycl. Polym. Sci. Technol.* **1**, 374 (1964).
61. "Acrylonitrile, Hazards, Properties, Handling, Storage," Publication PRT-107a, American Cyanamid Company, Chemical Products Division, Process Chemicals Department, Wayne, New Jersey (Received Feb. 19, 1990).
62. M. Imoto and K. Takemoto, *J.Polym. Sci.* **18**, 377 (1955); D. J. T. Hill, J. H. O'Donnell, and Paul W. O'Sullivan, *Macromolecules* **15**, 960 (1982).
63. L. Horner and W. Naumann, *Justus Liebigs Ann. Chem.* **587**, 93 (1954); S. Soennerskog, *Acta Chem. Scand.* **8**, 579 (1954).
64. K. Ziegler, W. Deparade, and H. Külhorn, *Justus Liebigs Ann. chem.* **567**, 151 (1950).
65. G. Salomon, *Rec. Trav. Chim. Pays-Bas* **68**, 903 (1949).
66. M. F. Shostakovskii and A. B. Bogdanova, *Zh. Prikl. Khim.* **24**, 495 (1951); *Chem. Abstr.* **46**, 1961 (1952).
67. Imperial Chemical Industries, Ltd., British Patent 567,778 (1945); *Chem. Abstr.* **41**, 2610 (1947); T. T. Jones and H. H. Melville, *Proc. Roy. Soc., Ser. A* **187**, 37 (1946).
68. J. A. Price, W. N. Thomas, and J. J. Padbury, U.S. Patent 2,626,946 (1953); *Chem. Abstr.* **47**, 670 (1953).

69. C. E. Schildknecht, "Vinyl and Related Polymers," pp. 270-271, Wiley, New York, 1952.
70. J. Bandrup and E. H. Immergut with W. McDowell, ed., "Polymer Handbook," 2nd ed., pp. IV-247, IV-248, J. Wiley and Sons, New York, 1975.
71. W. K. Wilkinson, U.S. Patent 3,087,919 (1963).
72. R. B. Parker and B. V. Moklar, U.S. Patent 3,161,511 (1964).
73. E. L. Kropa, U.S. Patents 2,356,767 (1944); 2,425,192 (1947).
74. Research Division of American Cyanamid Co., in "Polymers and Copolymers of Acrylonitrile," American Cyanamid Co., New York, 1956.
75. S. J. McCarthy, E. E. Elbing, I. R. Wilson, R. G. Gilbert, D. H. Napper, and D. F. Sangster, *Macromolecules* **19**, 2440 (1986); K. Kamide, Y. Miyazaki, and H. Kobayashi, *Polymer J.* **14** (8), 591 (1982).
76. Standard Oil of Ohio Research Laboratory, in "Methacrylonitric," Technical Bulletin Vistron Corporation, Cleveland, Ohio, 1968.
77. J. Furukawa, J. Tsuruta, and K. Morimoto, *Kogyo Kagaku Zasshi* **60**, 1402 (1957).
78. A. Zilkha, B. A. Feit, and M. Frankel, *J. Chem. Soc., London* p. 928 (1959).
79. M. Frankel, A. Ottolenghi, M. Albeck, and A. Zilkha, *J. Chem. Soc., London* p. 3858 (1959).
80. A. Zilkha, P. Neta, and M. Frankel, *J. Chem. Soc., London* p. 3357 (1960).
81. F. S. Dainton, D. M. Wiles, and A. N. Wright, *J. Polym. Sci.* **45**, 111 (1960); J. L. Down, J. Lewis, B. Moore, and J. Wilkinson, *Proc. Chem. Soc., London* p. 209 (1957).
82. R. B. Cundall, J. Driver, and D. D. Eley, *Proc. Chem. Soc., London* p. 170 (1958); R. B. Cundall, D. D. Eley, and R. Worrall, *J. Polym. Sci.* **58**, 869 (1962).
83. J. Ulbricht and R. Sourisseau, *Fraserforsch. Textiltech.* **12**, 547 (1961).
84. L. Horner, W. Jurgeleit, and K. Klüpfel, *Justus Liebigs Ann. Chem.* **591**, 108 (1955); D. C. Pepper, *Quart. Rev. Chem. Soc.* **8**, 88 (1954); N. S. Wooding and W. C. E. Higginson, *J. Chem. Soc., London* p. 774 (1952).
85. E. F. Evans, A. Goodman, and L. D. Grandine, in "Preparative Methods of Polymer Chemistry" (W. R. Sorenson and T. W. Campbell, eds.), 2nd ed., p. 283, Wiley (Interscience), New York, 1968.
86. J. R. Schaefgen, *Macromol. Syn.* **2**, 81 (1966).
87. "Glass Temperature Analyzer," Technical Bulletin SP-222, 3/65, Rohm and Haas Company, Philadelphia, Pennsylvania, 1965.
88. E. H. Riddle, "Monomeric Acrylic Esters," Van Nostrand-Reinhold, Princeton, New Jersey, 1954.
89. "Or-Chem Topics," No. 18. Rohm and Haas Company, Philadelphia, Pennsylvania, 1964.
90. "The Manufacture of Acrylic Polymers," Technical Bulletin SP-233, Rohm and Haas Company, Philadelphia, Pennsylvania, 1962.
91. J. O. Beattie, *Mod. Plast.* **33**, 109 (1956).
92. D. P. Hart, *Macromol. Syn.* **1**, 22 (1963).
93. H. P. Wiley, *J. Chem. Educ.* **25**, 204 (1948).
94. J. A. Simms, *J. Appl. Polym. Sci.* **5**, 58 (1961).
95. C. F. Ryan and J. J. Gormley, *Macromol. Syn.* **1**, 30 (1963).

96. K. J. Liu, J. S. Szuty, and R. Ullman, *Polym. Prepr., Amer. Chem. Soc., Div. Polym. Chem.* **5** (2), 761 (1964).
97. "2-Hydroxyethyl Methacrylate (HEMA) and Hydroxypropyl Methacrylate (HPMA)," Bulletin SP-216 8/64, Rohm and Haas Company, Philadelphia, Pennsylvania, 1964.
98. R. G. Bauer and N. C. Bletso, *Polym. Prepr., Amer. Chem. Soc., Div. Polym. Chem.* **10** (2), 632 (1969).
99. B. Wesslén and R. W. Lenz, *Polym. Prepr., Amer. Chem. Soc., Div. Polym. Chem.* **11** (1), 105 (1970).
100. R. Buter, Y. Y. Tan, and G. Challa, *J. Polym. Sci., Part A-1* **10**, 1031 (1972).
101. O. Wichterle and D. Lim, U.S. Patent 3,220,960 (1965).
102. F. W. Billmeyer, Jr., "Textbook of Polymer Science," 2nd ed., Wiley (Interscience), New York, 1971.
103. F. A. Bovey, I. M. Kolthoff, A. J. Medalia, and E. J. Meehan, "Emulsion Polymerization," Wiley (Interscience), New York, 1955.
104. E. W. Duck, *Encycl. Polym. Sci. Technol.* **5**, 801 (1966); J. G. Brodnyan, J. A. Cala, T. Konen, and E. L. Kelley, *J. Colloid Sci.* **18**, 73 (1963).
105. H. Fikentscher, H. Gerrens, and H. Schuller, *Angew. Chem.* **72**, 856 (1960).
106. H. Gerrens, *Fortschr. Hochpolym.-Forsch.* **1**, 234 (1959).
107. H. Gerrens, *Ber. Bunsenges. Phys. Chem.* **67**, 741 (1963).
108. W. D. Harkins, *J. Amer. Chem. Soc.* **69**, 1428 (1947).
109. G. Odian, "Principles of Polymerization," McGraw-Hill, New York, 1970.
110. "Emulsion Polymerization of Acrylic Monomers," Bulletin CM-104 A/cf, Rohm and Haas Company, Philadelphia, Pennsylvania.
111. W. V. Smith and R. H. Ewart, *J. Chem. Phys.* **16**, 592 (1948).
112. J. W. Vanderhoff, in "Vinyl Polymerization" (G. E. Ham *et al.*, eds.), Vol. 1, Part II, p. 1, Dekker, New York, 1969.
113. P. J. Flory, "Principles of Polymer Chemistry," Cornell Univ. Press, Ithaca, New York, 1953; M. S. Guillod and R. G. Bauer, *J. Appl. Polym. Sci.* **16**, 1457 (1972).
114. C. P. Roe, *Ind. Eng. Chem.* **60**, 20 (1968).
115. A. S. Dunn, *Chem. Ind.* (*London*) p. 1406 (1971).
116. W. S. Zimmt, *J. Appl. Polymer Science* **1**, 323 (1959).
117. H. W. Burgess, H. B. Hopfenberg, and V. T. Stannet, *Polym. Prepr., Amer. Chem. Soc., Div. Polym. Chem.* **10**, 1067 91969).
118. W. C. Mast and C. H. Fisher, *Ind. Eng. Chem.* **41**, 790 (1949).
119. W. E. Goode, R. P. Fellmann, and F. H. Owens, *Macromol. Syn.* **1**, 25 (1963).
120. C. Schuerch, W. Fowells, A. Yamada, F. A. Bovey, F. P. Hood, and E. W. Anderson, *Polym. Prepr., Amer. Chem. Soc., Div. Polym. Chem.* **5** (2), 1145 (1964).
121. N. S. Wooding and W. C. E. Higginson, *J. Chem. Soc., London* p. 774 (1952).
122. A. Zilkha, P. Neta, and M. Frankel, *J. Chem. Soc., London* p. 3357 (1960).
123. D. M. Wiles and S. Bywater, *Polym. Prepr., Amer. Chem. Soc., Div. Polym. Chem.* **4** (2), 317 (1963).
124. S. P. S. Yen and T. G. Fox, *Polym. Prepr., Amer. Chem. Soc., Div. Polym. Chem.* **10** (1), 1 (1969).
125. J. Trekoval and P. Kratochvil, *J. Polym. Sci., Part A-1* **10**, 1391 (1972).

126. L. S. Luskin and R. J. Meyers, *Encycl. Polym. Sci. Technol.* **1**, 246–328 (1964).
127. W. E. Goode, F. H. Owens, R. P. Fellmann, W. H. Snyder, and J. E. Moore, *J. Polym. Sci.* **46**, 317 (1960).
128. K. Hatada, M. Furomoto, T. Kitayama, Y. Tsubokura, and H. Yuki, *Polymer J.* **12**, 193 (1980); K. Hatada and T. Kitayama, Jpn. Kokai Tokkyo Koho JP 01,193,312 (89,193,312) Aug. 3, 1989; *Chem. Abstr.* **112**, 36750q (1990).
129. K. Hatada, T. Kitayama, S. Okahata, and H. Yuki, *Polymer J.* **14**, 971 (1982).
130. Author's Laboratory (WK).
131. O. Webster, U.S. Patent 4,417,034 (1983).
132. W. B. Farnham and D. Y. Sogah, U.S. Patent 4,414,372 (1983).
133. O. W. Webster, W. R. Hertler, D. Y. Sogah, W. B. Farnham, and T. V. Rajan Babu, *J. Am. Chem. Soc.* **105**, 5706 (1983).

11
AMINO RESINS

The condensation of compounds containing the amino groups with aldehydes, and in particular formaldehyde, gives materials commonly known as amino resins or aminoplasts. This section will concentrate on resins derived from urea and melamine. Formaldehyde reacts with the amino groups to give hydroxymethyl groups, which condense either with each other or with free amino groups to give resinous products. These products can be prepared to be either thermoplastic and soluble or crosslinked, insoluble infusible products known also as aminoplasts. Some other members of this class include formaldehyde condensation products of thiourea, ethyleneurea, guanamines, aniline, *p*-toluenesulfonamide, and acrylamide. Other aldehydes such as acetaldehyde, glyoxal, furfural, and acrolein have also been used in condensations with amino compounds to give resinous products.

In contrast to phenolic resins, the urea, melamine, and benzoguanamine–formaldehyde resins are colorless, odorless, and lightfast. Amino resins find use in decorative applications because of their clarity and colorless bonds. The arc resistance of amino resins makes them important components in electrical circuits. The amino resins, in contrast to phenolic resins, do not carbonize readily by the application of higher voltages between current-conducting parts and require 100 sec or more for development of an arc.

- **CAUTION:** The reaction of formaldehyde with hydrogen chloride has been shown to lead to the spontaneous production of the now known carcinogen *bis*(chloromethyl) ether. In addition, other aldehydes (crotonaldehyde, acetaldehyde, acrolein, etc.) are also toxic [1].

Although urea, melamine, and benzoguanamine are not considered highly toxic, little information is available on the toxicity of all their condensation products.

All other amino compounds should be considered suspect unless toxicity data are available. The use of efficient hoods and good personal hygiene (gloves, lab coats, and the like) are essential when amino resins are prepared.

1. UREA–FORMALDEHYDE CONDENSATIONS

Holzer [2] (1884) and Ludy [3] (1889) isolated a white precipitate by the reaction of urea with formaldehyde under acid conditions.

Goldschmidt in 1896 [4] studied the reaction of urea with formaldehyde in various strength acid solutions and obtained a granular white deposit of the empirical formula $C_5H_{10}N_4O_3$. In 1908, Einhorn and Hamburger [5] studied the same reaction in the presence of hydroxyl ions and, depending on the mole ratio of formaldehyde to urea, isolated mono- or dimethylolurea.

In 1927, Schiebler et al. [6] found that in acid solution the methylolureas are converted to insoluble substances similar to Goldschmidt's compound. Today the polymerization mechanisms involved are similar to those discussed for other methylol compounds such as phenolic resins or melamine resins. It is interesting to note that because urea has four active hydrogens and three sites for polymerization, linear, branched, and cyclic structures are possible. In fact, Kadowaki [7] has isolated several low molecular weight condensation products of urea–formaldehyde and has described their properties. The cyclic structures commonly called urones, such as dimethylolurone (N,N'-dimethyloltetrahydro-$4H$-1,3,5,-oxidiazin-4-one), have also been prepared by Kadowski.

11-1. Preparation of Urea–Formaldehyde Textile Resins (F : U Ratio 2 : 1) [8]

$$H_2NCONH_2 + 2CH_2O \longrightarrow HOCH_2NH\overset{\overset{\displaystyle O}{\|}}{C}NHCH_2OH \quad (1)$$

To a 3 liter, three-necked resin flask equipped with a mechanical stirrer, condenser, and thermometer are added 810 gm of 37% (10.0

moles) of formaldehyde, 4.0 gm of sodium acetate, and 8.0 gm of concentrated ammonia (28% NH_3). The mixture is stirred until all reactants have gone into solution and then 300 gm (5.0 moles) of urea is added. Heat is then applied to raise the temperature to 90°C over a 0.5 hr period; this temperature is maintained for 2 hr. Approximately 350 gm of water is removed under reduced pressure to yield a white turbid resin dispersion of 75% solids content with a pH of 5.5 and an acid number of less than 1.0. The pH is adjusted to 7.4 (with sodium hydroxide, sodium carbonate, or ammonium hydroxide) to stabilize the resin.

This resin is not only useful as a textile resin but also is of value in the manufacture of adhesives (with starch), wood glues for hot- and cold-pressed plywood, and in the paper wet strength resin area.

2. MELAMINE–ALDEHYDE CONDENSATIONS

Melamine was discovered by Liebig in 1834 and commercially produced [9] in the 1930s from dicyandiamide, but only with the publication of a patent by Soc. pour l'ind. chim. a'Bale [10] and one by Widmer and Fisch [11] were the reactions of melamine with formaldehyde described under a variety of conditions. Soon after, many publications and symposia appeared describing the latter reactions. Methylolmelamines undergo resinification reaction via esterification or self-condensation as in the case of the phenolic alcohols.

Melamine resin formation is therefore similar to phenolic resin production in that the A-stage resin is water soluble. The resin is partially dehydrated and the water solution that it forms is used to impregnate materials. Molding resins are filled with cellulose and impregnated in a vacuum mixer. The subsequent drying step converts the resin to the B stage, and it is then ground to a given particle size. Melamine–formaldehyde resins find use in many thermosetting resin applications such as molding resins; adhesives (mainly for plywood and furniture); laminating resins for counter, cabinet, and table tops; textile resins to impart crease resistance, stiffness, shrinkage control, water repellency, and fire retardance; wet-strength resins for paper; and in alkyd resin preparations to give baking enamels such as for automotive finishes.

A. FORMALDEHYDE CONDENSATIONS

Melamine reacts with a calculated amount of formaldehyde at pH 8.0 by heating rapidly to 70°–80°C for a few minutes, cooling to crystallize the products, washing the products with alcohol, and then air drying. Depending on the amount of formaldehyde used, the products are usually either di-, tri-, tetra-, penta-, or hexamethylolmelamines. The monomethylol compound has not been reported, probably because it is too unstable. All the methylol compounds are difficult to purify because of their solubility and instability toward heat. Hexamethylolmelamine is the most stable and easiest to purify. It can be prepared by heating for 10 min at 95°C a combination of melamine with 8 equivalents of formaldehyde. The product is isolated by cooling and filtering the white crystalline cake. Heating hexamethylolmelamine to 150°C converts it to an insoluble glass. Similarly, trimethylolmelamine can be prepared by stirring melamine with 3 moles of formaldehyde in the cold for 15–20 min. The methylol melamines are converted to the ethers by reaction with alcohols.

11-2. Preparation of Hexamethylolmelamine Prepolymer [12, 13]

$$\text{melamine} + 6CH_2=O \longrightarrow \text{hexamethylolmelamine} \quad (2)$$

To a resin flask equipped with a mechanical stirrer and condenser are added 37.8 gm (0.3 mole) of melamine and 195 gm of 37% (2.4 moles) formaldehyde solution. The pH of the mixture is adjusted with caustic to pH 8.3. The reaction mixture is heated on a water bath with stirring at 50°C for about 70–80 min, at which time all the melamine has gone into solution. On cooling the product separates as a thick white precipitate and is filtered to afford the monohydrate (after drying at 50°C). Yield: 62.5–69.0 gm (68–75%), m.p. 147°–153°C.

NOTE: Approximately one-third of the product melts at 120°–140°C.

REFERENCES

1. *Nat. Saf. News*, pp. 95–104 (1974).
2. H. Holzer, *Ber. Dtsch. Chem. Ges.* **17**, 659 (1884).
3. E. Ludy, *J. Chem. Soc.* **56**, 1059 (1889).
4. C. Goldschmidt, *Ber. Dtsch. Chem. Ges.* **29**, 2438 (1896).
5. A. Einhorn and A. Hamburger, *Ber. Dtsch. Chem. Ges.* **41**, 24 (1908).
6. H. Schiebler, F. Trosler, and E. Scholz, *Z. Angew. Chem.* **41**, 1305 (1928).
7. H. Kadowaki, *Bull. Chem. Soc. Jpn.* **11**, 248 (1936).
8. P. S. Hewett, U.S. Patent 2,456,191 (1948).
9. W. Fisch, U.S. Patent 2,164,705 (1939). I. G. Farbenindustrie A.-G., French Patent 817,895 (1937); G. Widmer, W. Fisch, and J. Jakl, U.S. Patent 2,161,940 (1939); D. Jayne, U.S. Patent 2,180,295 (1939); Gesellschaft für Chemische Industrie a Basel, French Patents 811,804 and 814,761 (1937).
10. Soc. pour l'ind. chim. a'Bale, British Patent 468,677 (1936).
11. G. Widmer and W. Fisch, U.S. Patent 2,197,357 (1940).
12. G. Widmer and W. Fisch, U.S. Patent 2,328,592 (1943).
13. S. R. Sandler, U.S. Patent 3,793,280 (1974).

12
PHENOL–ALDEHYDE CONDENSATIONS

In 1843, Pira [1] reported that phenol alcohols are converted to resins (called saliretins) on heating. Baeyer [2] in 1872 reported that the reaction of phenols with acetaldehyde in the presence of acid catalysts also gives resinous products. Kleeberg [3] in 1891 reported that formaldehyde undergoes similar reactions. However, Dianin [4, 5] found that acetone reacts with phenol to give a crystalline bisphenol (now known as bisphenol A). In 1874 Lederer [6] and Manasse [7] independently synthesized o-hydroxybenzyl alcohol (saligenin) by the low-temperature alkaline-catalyzed formaldehyde reaction.

Baekeland [8, 9] was granted a patent in 1909 describing his alkaline-catalyzed Bakelite resins ("resoles") and also the acid-cured "novolak" product. He conceived the idea of using counterpressure during hot cure to prevent bubbles and foaming from heat and was able to produce strong cured resinous products.

In connection with nomenclature, *Chemical Abstracts* uses the terms *novolak* and *resol*. Following more popular usage we use the terms *novolac* and *resole*.

- **CAUTION:** (*Toxicity*): Phenols, formaldehyde, and other aldehydes are toxic and should be handled with adequate ventilation and skin protection. Where possible the use of hydrogen chloride in the presence of formaldehyde or formaldehyde sources (hexamethylenetetramine and the like) should be avoided. Recent reports have indicated that formaldehyde and hydrogen chloride spontaneously react to give the known carcinogen *bis*(chloromethyl) ether.

12-1. Preparation of a Phenol–Formaldehyde Resole [10, 11]

[Reaction scheme: phenol + CH$_2$=O → oligomeric methylolated phenol structure with CH$_2$ bridges and CH$_2$OH groups]

(1)

(a) Method A [12]. To a resin kettle are added 800 gm (8.5 moles) of phenol, 80.0 gm of water, 940.5 gm (37%) of formaldehyde (11.6 moles), and 40 gm of barium hydroxide octahydrate. The reaction mixture is stirred and maintained at 70°C for 2 hr. Then sufficient oxalic acid is added to bring the pH to 6–7. The water is removed from the resin at 30–50 mm Hg at a temperature no higher than 70°C. Samples (1–2 ml) are withdrawn every 15 min to check the extent of condensation. The end point is taken when a sample of resin placed on a hot plate at 160°C gels in less than 10 sec or when the cooled resin is brittle and nontacky at room temperature. The resin at this point is termed resole or *A* stage. Further heating will convert it into a *B* stage and finally to a *C*-stage resin. The *A*-stage resin is used for adhesives, laminates, varnishes, and molding powders and is converted to the *C* stage by heating to 100°C for various times.

(b) Method B [13]. To a flask equipped with a reflux condenser and mechanical stirrer are added 47 gm (0.5 mole) of phenol, 80 ml of 37% aqueous formaldehyde (1.0 mole), and 100 ml of 4N sodium hydroxide. The reaction mixture is stirred at room temperature for 16 hr and then heated on a steam bath for 1 hr. The reaction mixture is cooled and the pH adjusted to 7.0. The aqueous layer is decanted from the viscous brown liquid product and the wet organic phase is taken up in 500 ml of acetone and dried over anhydrous MgSO$_4$ followed by molecular sieving. The dried acetone product solution is filtered and evaporated to yield a water-free light brown syrup. The IR spectrum of the uncured resole resin dried at room temperature for 3 days is shown in Fig. 1. The same resole resin cured for 3 hr at 120°C in air is shown in Fig. 2.

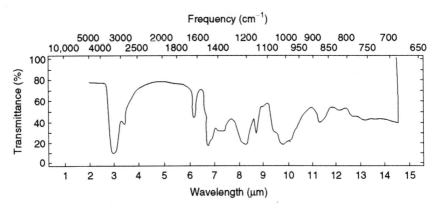

Figure 1 Infrared spectrum of uncured resole resin—3 days at room temperature. [Reprinted from W. M. Jackson and R. T. Conley, *J. Appl. Polym. Sci.* **8**, 2163 (1964). Copyright 1964 by the *Journal of Applied Polymer Science*. Reprinted by permission of the copyright owner.]

12-2. Preparation of a Phenol–Formaldehyde Casting Resin [14, 15]

(a) Method A [14]. To a resin kettle are added 100 gm (1.06 moles) of phenol, 200 gm of 37% formaldehyde (2.47 moles), and 12.0 gm of a 25% aqueous sodium hydroxide solution. The mixture is preheated for 25 min, refluxed for 17 min, neutralized, and then acidified to about pH 4.5 by the addition of approximately 19.5 gm of 51% aqueous lactic acid solution. The reaction mixture is dehy-

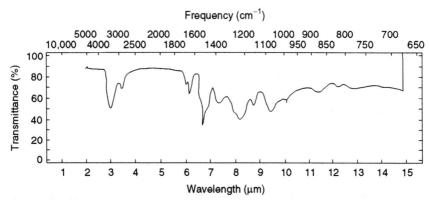

Figure 2 Infrared spectrum of a typical phenolic resin air-cured for 3 hr at 120°C. [Reprinted from W. M. Jackson and R. T. Conley, *J. Appl. Polym. Sci.* **8**, 2163 (1964). Copyright 1964 by the *Journal of Applied Polymer Science*. Reprinted by permission of the copyright owner.]

drated at 30-50 mm Hg at 80°C in order to remove the excess water. The dehydration takes 1.5-2.5 hr depending on the vacuum available. The final resin has about 14% water content. During the dehydration the resin is checked as in method A of Preparation 12-1 to determine when the product has reached the A stage. This resin, when poured into a suitable mold and heated at 75°-95°C in an oven, gives a clear, hard casting.

(b) Method B [15]. To a 1-liter resin kettle equipped with a mechanical stirrer, condenser, and thermometer are added 100 gm (1.06 moles) of phenol, 203 gm of 37% aqueous formaldehyde (2.5 moles), and 15 gm of 20% aqueous sodium hydroxide (0.075 mole). The reaction mixture is heated to 75°-80°C for about 2.5-3.5 hr. It is then vacuum-stripped of water at about 30 mm Hg while heating until the pot temperature gradually rises to about 64°C. At this point 7.5 gm (90% grade) of lactic acid (0.075 mole) is added to neutralize the starting alkali, followed by the addition of 15-20 gm of 98% glycerol. The vacuum removal of water is continued at low heat until the resin is almost completely dehydrated. When the resin is dehydrated sufficiently, a sample of the resin when dropped into a beaker of water of about 11°-13°C should congeal and solidify. The resin should also be hard enough only barely to yield when pressed between the fingers (**CAUTION:** Use gloves). The resin is now ready to be poured or cast into open molds and maintained at temperatures below the boiling point of water until completely set to a stable solid. Usually a period of 100-200 hr is required while heating at temperatures no higher than 78°-82°C.

Casting-type resins are generally prepared using a large excess of formaldehyde (2-3:1) with an alkali metal hydroxide as a catalyst. After reaction the pH is adjusted to about 4.5-6 with a weak organic acid. A polyol such as glycerol or ethylene glycol is sometimes added to improve the clarity of the product. The resin is cured by heating in open molds at 75°-95°C for extended periods of time. The casting is clear if sufficient water is removed during the resin preparation step.

REFERENCES

1. R. Pira, *Ann. Chem. Pharm.* **48**, 751 (1843).
2. A. Baeyer, *Ber. Dtsch. Chem. Ges.* **5** (1), 25-26, 280-282, and **5** (2), 1094-1100 (1872).

3. A. Kleeberg, *Justus Liebigs Chem. Ann.* **263**, 283 (1891).
4. A. Dianin, *J. Russ. Phys-Chem. Ges.* **1**, 488, 523, and 601 (1891).
5. A. Dianin, *Ber. Dtsch. Chem. Ges.* **25**, 334R (1892).
6. L. Lederer, *J. Prakt. Chem.* [N.S.] **50**, 223 (1894).
7. O. Manasse, *Ber. Dtsch. Chem. Ges.* **27**, 2409 (1894).
8. L. H. Baekeland, *J. Ind. Eng. Chem.* **1**, 149 (1909).
9. L. H. Baekeland, U.S. Patents 939,966 and 942,852 (1909); *J. Ind. Eng. Chem.* **6**, 506 (1913).
10. M. F. Drumm, *Am. Chem. Soc., Div. Org. Coat. Plast. Chem., Pap.* **26** 85 (1966).
11. A. Rudin, C. A. Fyfe, S. M. Vines, *J. Appl. Polym. Sci.* **28**, 2611 (1983).
12. *U.S. Dep. Commer., Off. Tech. Serv., PB Rep.* **25**, 642 (1945).
13. R. T. Conley and J. F. Bieron, *J. Appl. Polym. Sci.* **7**, 171 (1963).
14. R. W. Bentz and H. A. Neville, *J. Polym. Sci.* **4**, 671 (1949).
15. O. Pantke, U.S. Patent 1,909,786 (1933).

13
EPOXY RESINS

Epoxy resins are usually prepared from compounds (or polymers) containing two or more epoxy groups that have been reacted with amines, anhydrides, or other groups capable of opening the epoxy ring and forming thermosetting products. Polymers from monoepoxy compounds have been described in [1].

Schlack [2] and Castan [3, 4] are credited with the earliest U.S. patents describing epoxy resin technology. Greenlee [5] further emphasized the use of bisphenols and their reaction with epichlorohydrin to yield diepoxides capable of reaction with crude tall oil resin acids to yield resins useful for coatings. The use of diepoxide resins that are cured with amines was reported by Whittier and Lawn [6] in a U.S. patent in 1956.

The introduction of epoxidation techniques for polyunsaturated natural oils by Swern [7, 8] led to industrial interest in the preparation of epoxy compounds useful for resin production [9, 10].

Epoxy resin technology has been reviewed and a number of relevant references are available [7, 11–14].

Safety precautions. Epoxy resins and their curing agents are considered primary skin irritants. Contact with epoxy resins should only be made using gloves and face shields and while working in hoods or well-ventilated areas [15]. Some individuals on prolonged contact with epoxy resins may develop a skin sensitization evidenced by blisters or other dermatitis conditions. Other individuals may develop an asthma-like condition. Contaminated gloves and work clothes should be changed immediately and either laundered or discarded. Contaminated shoes should be discarded. Frequent washing of hands is advisable and strict personal hygiene must be practiced. Note that some aromatic amine curing agents may be carcinogenic. A properly cured epoxy resin system usually presents no health problems relating to skin irritation. Individuals who show sensitivity should discontinue handling epoxy compounds. Otherwise hypersensitivity develops, making it difficult even to come close to those materials without developing dermatitis or other reactions. References [14] and [16] should be referred to for additional safety information.

1. ANALYSIS OF EPOXY RESINS

Epoxy resins are analyzed for epoxy or oxirane content, which is reported as the epoxy or oxirane equivalent or epoxy equivalent weight, i.e., the weight of resin in grams that contains one gram equivalent of an epoxy group. The "epoxy value" designates the fractional number of epoxy groups per 100 gm of epoxy resin. Percent oxirane oxygen is used for epoxidized oils and dienes.

Analytically, epoxy groups are determined by the reaction with hydrogen halide and back titration with standard base. Other functional groups present may cause interference problems and result in poor end points. Pyridinium chloride–pyridine is a recommended reagent for the analysis of bisphenol–diglycidyl ether resins [17, 18].

Another improved analytical procedures allow direct titration of the epoxy group [18]. This is achieved by the use of hydrogen bromide dissolved in an anhydrous protic solvent such as glacial acetic acid [19]. The methods developed by Jay [20] and Dijkstra and Dahmen [21] using quaternary ammonium bromide or iodide in acetic and perchloric acid solutions for titration are considered the best general techniques for a wide variety of epoxides (hindered and unhindered alicyclic epoxides). The use of the halogen acid procedure fails for epoxides that undergo intramolecular rearrangement to aldehydes or ketones [22].

More recently Eggers and Humphrey have reported on the application of gel permeation chromatography to monitor epoxy resin molecular distributions and curing [23].

2. CURING-POLYMERIZATION REACTIONS OF EPOXY COMPOUNDS AND RESINS

Epoxy resins are cured [24] by reaction of the epoxy group with other functional groups to give linear, branched, or crosslinked products as described in Eq. (1):

$$Z + CH_2\underset{O}{-}CHR \longrightarrow Z^+ - CH_2 - \underset{O^-}{CH} - R \quad (1)$$

$$Z = R_3N, ROH, RCOOH, (RCO)_2O, RNH_2, RCONH_2, RSH, \text{etc.}$$

The compounds Z are reactive compounds such as amines anhydrides, and acids [25]. The curing reaction can also involve homopolymerization catalyzed by Lewis acids or tertiary amines.

With diglycidyl derivative of bisphenol A, the aromatic amines such as 4,4'-methylene dianiline or diaminodiphenyl sulfone provide good thermal stability for the final cured resin. Although the aliphatic primary amines react more rapidly (triethylenetetramine cures the above epoxy resin based on bisphenol A in 30 min at room temperature and causes it to exotherm up to 200°C), they are more difficult to handle and offer poor thermal stability.

The anhydrides generally provide pot lives of days or months and usually cure at 100°–180°C with very little exotherm. Tertiary amines accelerate the time for gelation but still require the elevated temperature cure to obtain optimum properties. Anhydrides usually give brittle products but the addition of polyether flexibilizing groups yields more elastomeric products [26].

13-1. Curing Reaction of Bisphenol A Diglycidyl Ether with Poly(adipic acid anhydride) [3]

Three hundred grams of the epoxy resin is mixed with 125 gm of poly(adipic acid anhydride) and heated to 130°C for 1 hr. The resulting resin hardens at 150°C in 1.5 hr and is rather elastic. If the resin is heated at 150°–170°C for any length of time it becomes crosslinked and shows good adhesion to a wide variety of nonporous materials (glass, metal).

Curing reactions of epoxy compositions are described in Tables I and II.

TABLE I
Curing Reactions of the Diglycidyl Ether of Bisphenol A

Curing agents	Temp. (°C)	Ref.
Phthalic anhydride	120–170	a
Polyadipic acid anhydride	130	a
Coal tar pitch and diethylene triamine	—	b
Tall oil plus cobalt naphthenate	—	c
Organosiloxanes containing Si–OH groups, plus catalysts [ferric naphthenate, ferric bromide, zinc acetate, zinc benzoate, zinc stearate]	—	d
Triethylenetetramine plus molasses	—	e

[a] P. Castan, U.S. Patent 2,324,483 (1943).
[b] F. Whittier and R. J. Lawn, U.S. Patent 2,765,288 (1956).
[c] S. O. Greenlee, U.S. Patent 2,493,486 (1950).
[d] C. L. Frye and W. M. McLean, U.S. Patent 3,055,858 (1962).
[e] T. Ramos, U.S. Patent 2,894,920 (1959).

TABLE II
Curing Reactions of Various Epoxy Compounds and Resins

Epoxy compound or resins

$$-RO-CH_2-\overset{O}{\overset{\diagup \diagdown}{CH-CH_2}}$$

R equals or is derived from	Curing agents	(°C)[a]	Ref.
Bisphenol A	Triethylene tetraamine	–(234°F)	b
Bisphenol A	Ester dianhydride derived from trimellitic anhydride	100 (1 hr)	c
Novolac resin	4,4-Methylenedianiline	100 (2 hr)	c
Epon 562	4-p-Phenyl trimellitate anhydride	125	d
Dinaphthol (4,4'-dihydroxy-1,1'-dinaphthyl)	Diethylenetetramine	105 (18 hr)	e
	Maleic anhydride + trace $(CH_3)_2N-C_6H_5$	160 (1 hr)	e
	Phthalic anhydride	160 (80 hr)	e
Epikote 828	Condensates of trimellitic acid anhydride with liquid hydrogenated polybutadiene or butadiene polymer containing $\geq$ 1 OH group + trace benzylmethylamine	90–130	f
Glycidyl ether epoxy resin	Acid anhydride + trace of organic phosphonium salt	—	g

TABLE II (Continued)
CURING REACTIONS OF VARIOUS EPOXY COMPOUNDS AND RESINS

Epoxy compound or resins

$$-RO-CH_2-\overset{O}{\overset{\diagup\diagdown}{CH-CH_2}}$$

R equals or is derived from	Curing agents	(°C)[a]	Ref.
Bisphenol	Polysebacic polyanhydride	90	h
	Phthalic anhydride–triethylamine	—	i
	BF_3–monoethylamine	—	j
	Stannous octoate alone or with amines, anhydrides, polycarboxylic acids, polythiols, polyisocyanates, polyhydric phenols	—	k
	Amides and polyamides	—	l
	Ureas	—	m
	Polyurethanes	—	n
	Isocyanates		

[a] Times given in parentheses indicate total time at a given temperature required to fully cure the epoxy resin.
[b] T. Ramos, U.S. Patent 2,894,920 (1959).
[c] S. R. Sandler and F. R. Berg, U.S. Patent 3,437,671 (1969).
[d] D. F. Loncrini, U.S. Patent 3,140,299 (1964).
[e] V. C. E. Burnop, *J. Appl. Polym. Sci.* **12**, 699 (1968).
[f] A. Fukami and T. Moriwaki, Japan Kokai 73/88,198 (1973).
[g] J. D. B. Smith, U.S. Patent 3,784,583 (1974).
[h] R. G. Black, J. J. Seiwert, and J. B. Boylan, *Plast. Technol.* **10**, 37 (1964).
[i] K. Cressy and J. Delmonte, *136th Am. Chem. Soc. Meet., Atlantic City, N.J., 1959* Paper 30, p. 100 (1959).
[j] S. O. Greenlee, U.S. Patent 2,521,912 (1950).
[k] W. R. Proops and G. W. Fowler, U.S. Patent 3,117,099 (1964).
[l] S. O. Greenlee, U.S. Patent 2,589,245 (1952).
[m] S. O. Greenlee, U.S. Patent 2,713,569 (1955).
[n] Olin Matheson Corp., Netherlands Patent Appl. 6,411,810 (1965).

REFERENCES

1. S. R. Sandler and W. Karo, "Polymer Syntheses," 2nd ed., Vol. 1, Academic Press, Boston, 1992.
2. P. Schlack, U.S. Patent 2,131,120 (1938); German Patent 676,117 (1939).
3. P. Castan, U.S. Patent 2,324,483 (1943).
4. P. Castan, Swiss Patent 211,114 (1940).
5. S. O. Greenlee, U.S. Patent 2,493,486 (1950).
6. F. Whittier and R. J. Lawn, U.S. Patent 2,765,288 (1956).

7. D. Swern, *Chem. Rev.* **45**, 1 (1949).
8. T. W. Findley, D. Swern, and J. T. Scanlan, *J. Am. Chem. Soc.* **67**, 412 (1945).
9. B. Phillips, "Peracetic Acid and Derivatives," Bulletin P-58-0283, Union Carbide Chemical Co., New York, 1958; Bulletin 4, Food Machinery and Chemical Corp., Becco Chem. Div., 1952 (revised 1957); R. J. Gall and F. P. Greenspan, *Ind. Eng. Chem.* **47**, 147 (1955); Bulletin P61-454, E. I. Du Pont de Nemours & Co., 1954; Bulletin A6282, E. I. Du Pont de Nemours & Co., 1955.
10. E. P. Greenspan and R. J. Gall, *J. Am. Oil Chem. Assoc.* **33**, 391 (1956).
11. S. R. Sandler and W. Karo,"Organic Functional Group Preparations," Vol. 1, pp. 99–115, Academic Press, New York, 1968.
12. H. Lee and K. Neville, "Epoxy Resins, Their Applications and Technology," McGraw-Hill, New York, 1957; Handbook of Epoxy Resins, McGraw-Hill, New York, 1967; *Encycl. Polym. Sci. Technol.* **6**, 209 (1967); I. Skeist, "Epoxy Resins," Van Nostrand-Rheinhold, Princeton, New Jersey, 1958; M. W. Ranney, "Epoxy and Urethane Adhesives," Noyes Data Corp., New Jersey, 1971; G. R. Somerville and P. D. Jones, *Abstr. 168th Am. Chem. Soc. Meet.*, Atlantic City, New Jersey, 1974, Abstract ORPL, 146 in the Organic Coatings and Polymer Chemistry Division (1974).
13. L. V. McAdams and J. A. Gannon, *Encycl. Polym. Sci. Eng.* (2nd ed.) **6**, 322 (1986).
14. F. Lohse, *Makromal. Chem., Macromol. Symp.* **7** (1987); *Chem. Abstr.* **106**, 103051d (1987).
15. "Industrial Hygiene Bulletin on Handling Epon Resins and Auxiliary Chemicals," SC 62-33, Shell Chem. Corp., 1962.
16. J. E. Berger, K. I. Darmer, and E. F. Phillips, *High Solids Coat.* **16** (Dec. 1980).
17. D. W. Knoll, D. H. Nelson, and P. W. Keheres, *Pap., 134th Am. Chem. Soc. Meet.*, Chicago, 1958, Division of Paint, Plastics, and Printing Ink Chemistry, Paper No. 5, p. 20 (1958).
18. B. Dobinson, W. Hoffmann, and B. P. Stark, "The Determination of Epoxide Groups," Pergamon, Oxford, 1969.
19. A. J. Durbetaki, *Anal. Chem.* **28**, 2000 (1956).
20. R. R. Jay, *Anal. Chem.* **36**, 667 (1964).
21. R. Dijkstra and E. A. M. F. Dahmen, *Anal. Chim. Acta* **31**, 38 (1964).
22. A. J. Durbetaki, *Anal. Chem.* **29**, 1666 (1957).
23. E. A. Eggers and J. S. Humphrey, Jr., *J. Chromatogr.* **53**, 33 (1971).
24. H. Lee and K. Neville, "Handbook of Epoxy Resins," Chapter 5, McGraw-Hill, New York, 1967.
25. "Epon Resins," Technical Brochure SC 71-1, Shell Chem. Corp., Houston, Texas, 1971.
26. S. R. Sandler and F. R. Berg, U.S. Patent 3,437,671 (1969).

14
ALKYD RESINS

Alkyd resins are prepared from polyhydric alcohols (three or more hydroxyls), polybasic acids, and monobasic fatty acids (saturated and unsaturated) [1]. They are soluble in hydrocarbon solvents and are used in the coating industry. Polyesters prepared from dihydric alcohols are not included in the area of alkyd resins in this chapter and have been described in [2].

Alkyd-type resins were first described in 1847 by Berzelius [3], who obtained a brittle resinous polymer by the reaction of tartaric acid and glycerol. In 1853 Berthelot prepared the glycerol ester of camphoric acid [4]. In 1856 von Bemmelen [5] studied resins obtained by heating glycerol with succinic acid, citric acid, and a mixture of succinic and benzoic acids. Other early investigators were Dubus [6], Lourenco [7], and Furaro and Danesi [8]. In 1901 Smith [9] described a solid transparent glycerol phthalate resin that was insoluble in water but soluble in hot glycerol. Removal of the glycerol *in vacuo* afforded a resinous mass containing glycerol and phthalic anhydride in a molecular ratio of approximately 2:3. Further heating causes the resinous material to yield a puffy brittle mass. General Electric Laboratories carried out a series of intensive investigations on the glycerol–phthalic anhydride reactions leading to the production of useful resins of the heat-convertible type as summarized in the work of Callahan [10], Friberg [11], Arsem [12], Dawson [13], and Howell [14]. The latter workers found that replacing some of the phthalic anhydride with monobasic acids leads to more flexible resins with better solubility properties. Kienle [15, 16] in 1929 and 1933 prepared glycerol–phthalate resins containing some unsaturated fatty acids; these came into use as protective coatings that were cured by air drying or heating at low baking temperatures. The commercial production [17] of phthalic anhydride by the catalytic oxidation of naphthalene accelerated progress in commercially manufactured alkyd resins.

Some of the typically used raw materials are shown in Table I.

The equipment used is either stainless steel or stainless steel clad and is corrosion resistant. An example of a typical equipment setup is shown in Fig. 1, where the sparge is either nitrogen or carbon dioxide.

TABLE I
REPRESENTATIVE ALKYD RESIN STARTING MATERIALS

Raw material	m.p. (°F)	b.p. (°F)	Mol. wt.	Sapon. no.	Equiv. wt.
Polyhydric compounds					
Glycerin	—	554	92.1	—	31
Ethylene glycol	—	386	62.1	—	31
Diethylene glycol	—	473	106.1	—	53
Propylene glycol	—	374	76.1	—	38
Trimethylolethane	395	—	120.1	—	40
Trimethylolpropane	136	—	134.2	—	44.7
Pentaerythritol	504	—	136.2	—	45.5
Acids–Anhydrides[a]					
Phthalic anhydride	270	544	148.1	—	74
Phthalic acid	375–410	—	166.1	—	83
Isophthalic acid	650	—	166.1	—	83
Terephthalic acid	Sublimes 570	—	166.1	—	83
Trimellitic anhydride	334	—	192.1	—	64
Trimellitic acid	419–423	—	210.1	—	70
Maleic anhydride	266	—	116.1	—	49
Fumaric acid	548	—	116.1	—	49
Adipic acid	306	—	146.1	—	73
Oils					
Linseed	—	—	—	188–196	280
Soya	—	—	—	189–195	280
Dehydrated castor	—	—	—	188–194	293
Safflower	—	—	—	188–194	293
Tall	—	—	—	170–180	285

[a]**Caution:** Operations with all acids or anhydrides should be carried out in a well-ventilated hood using proper personal protection (gloves, apron, chemical goggles).

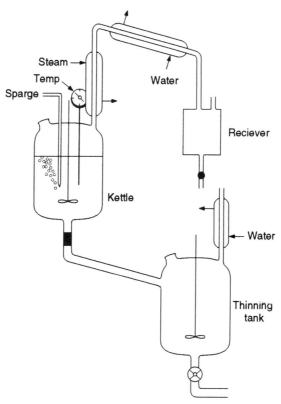

Figure 1 Alkyd resin processing equipment. [Reprinted from "Amoco TMA as Primers for Alkyd-Melamine Enamels and Acrylic Lacquers," Technical Bulletin TMA 25a, Amoco Chemical Corp., Chicago, Illinois, 1974.]

The earlier alkyd resins (glyceryl phthalate) were all of the heat-convertible variety; i.e., on heating the resin was converted to a nonfusible type [15]. More recently an unmodified alkyd refers to a resin prepared from a dibasic acid, a polyol, and an oil or fatty acid. In some cases a monofunctional acid such as benzoic acid has been added to reduce functionality [18]. Oxygen- or air-convertible alkyd resins are prepared by replacing part of the polybasic acid with the proper amount of an oxidizable unsubstituted fatty acid or acids such as linolenic, oleostearic, or other acids that are commercially available. With these two types of systems, alkyds are used to prepare coatings that cure upon baking or air drying.

14-1. Preparation of Poly(glyceryl phthalate) Alkyd Resin [19]

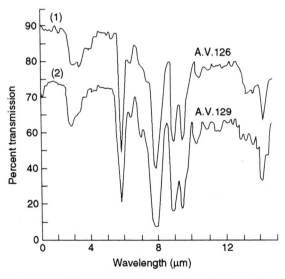

Figure 2 Infrared spectra: (1) glyceryl phthalate (acid), A.V. = 129; (2) glyceryl phthalate (anhydride), A.V. = 126. [Reprinted from R. H. Kienle, P. A. Van Der Menlen, and F. E. Petke, *J. Am. Chem. Soc.* **61**, 2268 (1939). Copyright 1939 by the American Chemical Society. Reprinted by permission of the copyright owner.] *Note:* A.V. = acid value.

To a resin flask equipped with a thermometer, mechanical stirrer, condenser, Dean & Stark trap, and nitrogen bubbler are added 46.0 gm (0.50 moles) of glycerol and 111 gm (0.75 moles) of phthalic anhydride. The reaction mixture is heated to 195°C for 1.5 hr while a slow stream of nitrogen is bubbled through it. The acid value is about 170 at this point, and the product is still soluble in acetic acid, acetone, and other solvents. Further heating at 195°C for another 1.5 hr causes the resin to set to an insoluble gel (acid no. = 119–136). The infrared spectra are shown in Fig. 2. The reaction data are shown in Table II along with percent esterification data in Table III and Fig. 3. The Dean & Stark trap in the above preparation contained 7.0 gm of a liquid consisting of water and some white crystals. Titration with $N/25$ potassium hydroxide using phenolphthalein as an indicator showed 0.085 gm of phthalic anhydride (0.076%).

TABLE II
ISOTHERMS OF FREE ACIDITY CHANGE WITH TIME AT 195°C[a]

Time (sec)	Time (min)	Free anhydride (%)	Acid number
0	0	70.67	535
215	3.58	31.05	235
720	12.16	28.99	218
1320	23.0	27.02	212
2440	40.67	25.25	191
3420	50.0	24.51	185
3980	63.0	23.96	181
4380	73.0	23.42	177
5790	96.5	22.23	169
6780	113.0	21.47	163
8340	139.0	21.02	159
9000	150.0	19.93	151
9720	162.0	19.84	149
10,610	177.0	19.18	145
11,150	186.0	18.31	139
11,340	189.0	17.94	136
11,700	195.0	15.81	119

[a] Reprinted from R. H. Kienle and A. G. Hovey, *J. Am. Chem. Soc.* **51**, 509 (1929). Copyright 1929 by the American Chemical Society. Reprinted by permission of the copyright owner.

TABLE III
Variation of Properties of Products with Time at 195°C[a]

Time (sec)	Time (min)	Flow pt. (°C)	Refr. index	Acid number	Sapon. number	Ester value	Ester (%)
0	0	—	—	(535)	—	0	0.0
215	3.58	64	—	235	584	349	59.8
720	12.16	69	—	218	582	364	62.5
1380	23.0	76	1.56	212	583	371	63.6
2440	40.67	78	—	191	570	379	66.5
3420	50.0	81	—	185	589	404	68.5
3980	63.0	82.5	—	181	591	410	69.4
4380	73.0	83	—	177	588	411	69.9
5740	96.5	86	—	169	589	420	71.4
6780	113.0	91	—	163	605	442	73.0
8340	139.0	96	—	159	611	452	73.8
9000	150.0	96	—	151	604	453	75.0
9720	162.0	105	—	149	605	456	75.5
10,610	177.0	106	—	145	608	463	76.1
11,150	186.0	118	1.58	139	614	475	77.3
11,340	189.0	Gel	—	136	585	449	77.8
11,700	195.0	Gel	—	119	574	455	79.3

[a] Reprinted from R. H. Kienle and A. G. Hovey, *J. Am. Chem. Soc.* **51**, 509 (1929). Copyright 1929 by the American Chemical Society. Reprinted by permission of the copyright owner.

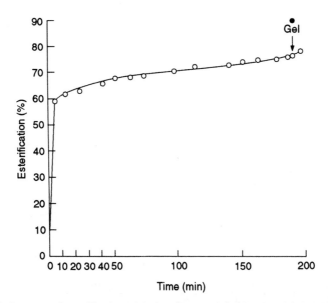

Figure 3 Increase of esterification with time ($T = 195°C$). [Reprinted from N. H. Kienle and A. G. Hovey, *J. Am. Chem. Soc.* **51**, 509 (1929). Copyright 1929 by the American Chemical Society. Reprinted by permission of the copyright owner.]

REFERENCES

1. R. H. Kienle and A. G. Hovey, *J. Am. Chem. Soc.* **51**, 509 (1929).
2. S. R. Sandler and W. Karo, "Polymer Syntheses," 2nd ed., Vol. 1, pp. 68–86, Academic Press, Boston, 1992.
3. J. Berzelius, *Rapp. Annu. Inst. Geol. Hong.* **26**, 1 (1847).
4. M. M. Berthelot, *C. R. Hebd. Seances Acad. Sci.* **37**, 398 (1853).
5. J. von Bemmelen, *J. Prakt. Chem.* **69**, 84 and 93 (1856).
6. H. Dubus, *Philos. Mag.* [4] **16**, 438 (1858); *Jahrb. Fortschr. Chem.* p. 431 (1856).
7. A. V. Lourenco, *Ann. Chim. Phys.* [3] **7**, 67 and 313 (1863).
8. A. Furaro and F. Danesi, *Gazz. Chim. Ital.* **10**, 56 (1880); *Jahrb. Forstchr. Chem.* p. 799 (1980).
9. W. Smith, *J. Soc. Chem. Ind., London* **20**, 1073 (1901).
10. M. J. Callahan, U.S. Patents 1,108,329; 1,108,330; 1,108,332; 1,091,627; 1,091,628; and 1,091,732 (1914).
11. L. H. Friberg, U.S. Patent 1,119,592 (1914).
12. W. C. Arsem, U.S. Patents 1,098,776 and 1,097,777 (1914).
13. E. S. Dawson, U.S. Patent 1,141,944 (1915).
14. K. B. Howell, U.S. Patent 1,098,728 (1914).
15. R. H. Kienle and C. S. Ferguson, *Ind. Eng. Chem.* **21**, 349 (1929).
16. R. H. Kienle, U.S. Patent 1,893,873 (1933).
17. H. D. Gibbs, *J. Ind. Eng. Chem.* **11**, 1031 (1919).
18. C. R. Martens, "Alkyd Resins," pp. 51–59, Van Nostrand-Reinhold, Princeton, New Jersey, 1961.
19. R. H. Kienle and A. G. Hovey, *J. Am. Chem. Soc.* **51**, 509 (1929); Anonymous, *Bull. Union Physicians* **83** (714), 662 (1989); *Chem. Abstr.* **112**, 54392d (1989).

15
POLYACETALS AND POLY(VINYL ACETALS)

Polyols and polythiols may be reacted with aldehydes or ketones to form polyacetals, polythioacetals, polyketals, and polythioketals, respectively. The preparations are not confined to the reactions of simple carbonyl reagents with simple, linear diols. These may give rise to linear polymers. Polyols such as pentaerythritol may be reacted to form cyclic acetals such as polyspiroacetals. The preparation of polyketals has not been studied as extensively as the preparation of polyacetals. The reactions of ketones with polyols are more sluggish than those with the corresponding aldehydes.

Of particular importance are the reaction products of the various grades of poly(vinyl alcohol) with carbonyl compounds such as formaldehyde, acetaldehyde, and butyraldehyde. These products are classified as poly(vinyl acetals). Poly(vinyl butyral) has been used extensively in the manufacture of automobile safety glass. Essentially, such safety glasses are produced by adhering an interlayer film of the resin between two sheets of high-quality glass.

The preparation of acetals and ketals by the reaction of a carbonyl compound with polyols are summarized in Eqs. (1) and (2). The formation of a polyspiroacetal resin is shown in Eq. (3) [1–5].

$$n\text{RCH}=\text{O} + n\text{HX}-\text{R}'-\text{XH} \xrightarrow{n\text{H}_2\text{O}} \left[\text{R}'-\text{X}-\underset{\underset{\text{R}}{|}}{\text{CH}}-\text{X} \right]_n \quad (1)$$

Based on S. R. Sandler and W. Karo, "Polymer Syntheses," 2nd ed., Vol. II, pp. 192ff, by permission of Academic Press, San Diego, 1994.

$$n\text{R}_2\text{C}=\text{O} + n\text{HX}-\text{R}'-\text{XH} \xrightarrow{n\text{H}_2\text{O}} \left[\text{R}'-\text{X}-\underset{\underset{\text{R}}{|}}{\overset{\overset{\text{R}}{|}}{\text{C}}}-\text{X} \right]_n \quad (2)$$

X = O or S

$$\underset{\text{HOCH}_2}{\overset{\text{HOCH}_2}{>}}\text{C}\underset{\text{CH}_2\text{OH}}{\overset{\text{CH}_2\text{OH}}{<}} + \text{R(CHO)}_2 \longrightarrow \left[\underset{\text{O}}{\overset{\text{O}}{>}}\!\!\!\!\!\bowtie\!\!\!\!\!\underset{\text{O}}{\overset{\text{O}}{<}}\!\!-\text{R} \right]_n \quad (3)$$

The acetalization of poly(vinyl alcohol) is illustrated in a generalized manner in Eq. (4). Note that not only do cyclic acetals form with the hydroxyl groups that are 1,3- to each other on the same polymer chain, but crosslinks may form with hydroxyls on adjacent chains. To add to the complexity of the formation of poly(vinyl acetals), it must be kept in mind that the base polymer consists of material with a distribution of molecular weights and with its own varieties of crosslinks and other structural features. Nevertheless, poly(vinyl acetals) are commercially produced and utilized.

$$\left[\text{CH}_2-\underset{\underset{\text{OH}}{|}}{\text{CH}}-\text{CH}_2-\underset{\underset{\text{OH}}{|}}{\text{CH}} \right]_n + n\text{RCH}=\text{O}$$

$$\text{H}^+ \Big\downarrow -\text{H}_2\text{O}$$

$$\left[\text{CH}_2-\text{CH} \underset{\text{O}}{\overset{\text{CH}_2}{<}} \text{CH}-\text{CH}_2 \atop \underset{\underset{\text{R}}{|}}{\text{CH}} \right]_n + \text{some } \sim\!\!\text{CH}_2-\underset{\underset{\text{OH}}{|}}{\text{CH}}-\text{CH}_2-\underset{\underset{\text{O}}{|}}{\text{CH}}\!\!\sim \quad (4)$$

$$\underset{\underset{\text{OH}}{|}}{\overset{\overset{\text{CHR}}{|}}{\sim\!\!\text{CH}_2-\text{CH}-\text{CH}_2-\text{CH}\!\!\sim}}$$

R = H, alkyl, or aryl

Carothers was one of the first to exploit the polyacetal-forming reaction in his attempt to prepare polyformals by the reaction of formaldehyde [6] with glycols. He found that diols below 1,5-pentanediol led to cyclic formals, whereas the principal product from 1,6-hexanediol and from higher molecular weight diols was a polyformal [6]. Similar products were formed by the acetal interchange reaction when appropriate diols were used (Eq. 5) [7].

$$n\text{ROCH}_2\text{OR} + n\text{HOR'OH} \xrightarrow[\text{acid catalyst}]{-2n\text{ROH}} \text{+R'}-\text{OCH}_2-\text{O+}_n \qquad (5)$$

Polyacetals and polyketals can be prepared by most of the methods shown in Scheme 1 and Scheme 2 starting with diols, polyols, or polythiols in place of the alcohols [8].

SCHEME 1
PREPARATION OF ACETALS [8]

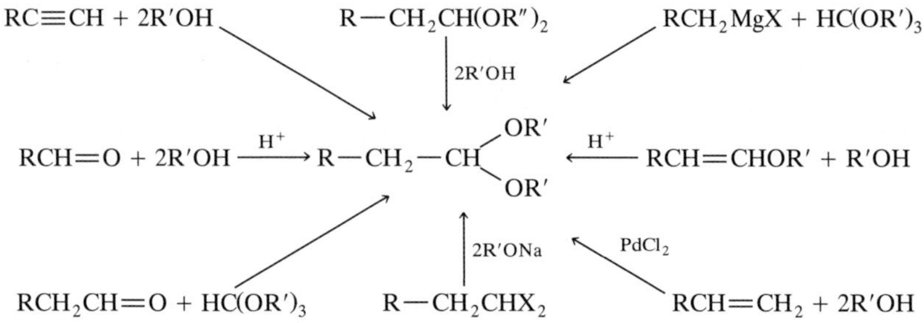

```
RC≡CH + 2R'OH              R—CH₂CH(OR")₂              RCH₂MgX + HC(OR')₃
                                   │
                                 2R'OH
                                   
                                        OR'
RCH=O + 2R'OH  ──H⁺──→  R—CH₂—CH          ←──H⁺──  RCH=CHOR' + R'OH
                                        OR'

                                   │              PdCl₂
                                 2R'ONa
RCH₂CH=O + HC(OR')₃        R—CH₂CHX₂                   RCH=CH₂ + 2R'OH
```

SCHEME 2
PREPARATION OF KETALS [8]

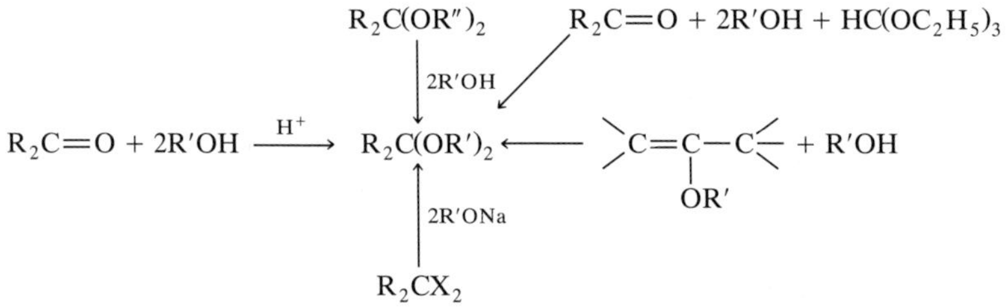

```
                           R₂C(OR")₂         R₂C=O + 2R'OH + HC(OC₂H₅)₃
                                │
                              2R'OH
R₂C=O + 2R'OH  ──H⁺──→  R₂C(OR')₂  ←──  >C=C—C< + R'OH
                                                │
                                │              OR'
                              2R'ONa
                           R₂CX₂
```

Another method for the synthesis of polyacetals involves *transacetalization* reactions of diols with acetylenic compounds [9] or vinyl ethers [10].

Polyacetals, as distinct from polyformaldehyde, have not been exploited to their fullest potential as other polymers have [11].

In part, the interest in the polyacetals arises from a controversy about the reaction conditions and ratio of reactants, which may lead either to macro-cyclic or linear polymeric products. There is also the general interest in intermediate spiro-compounds that may form, for example by the reaction of diols with dialdehydes. Such compounds might expand, rather than contract. In addition to Refs. [7] and [11] some reviews can be found in [12–18].

1. POLYACETALS FROM FORMALDEHYDE

15-1. Preparation of the Polyformal of Diethylene Glycol [19]

$$HOCH_2CH_2OCH_2CH_2OH + CH_2=O \longrightarrow$$

$$H\text{+}OCH_2CH_2O-CH_2CH_2OCH_2\text{+}_n OH \quad (6)$$

To a three-necked round-bottomed flask equipped with a mechanical stirrer, condenser, and Dean & Stark trap are added 106 gm (1.0 mole) of diethylene glycol, 33 gm (1.0 mole) of 91% paraformaldehyde, 0.1 ml of concentrated sulfuric acid, and 20 gm of toluene. The mixture is heated under reflux; 18 gm (4.0 moles) of water is collected. The toluene is removed under reduced pressure (20 mm Hg) and heating is continued to 150°C. The resulting product is neutralized with dilute sodium hydroxide to pH 7. The polymer is soluble in water and toluene and has a molecular weight of 480. The product has an OH equivalent of 220 and n_D^{30} of 1.462. The product is suggested as useful as a textile finishing agent for paper, cotton [20], and leather.

2. POLYACETALS FROM SUBSTITUTED ALDEHYDES

The use of an aldehyde higher than formaldehyde, such as acetaldehyde or benzaldehyde, affords polyacetals containing a side group, as shown in Eq. (7):

$$HO(CH_2)_n OH + RCH=O \longrightarrow \left[\begin{matrix} (CH_2)_n O-CH-O \\ | \\ R \end{matrix} \right]_n + H_2O \quad (7)$$

R = alkyl, aryl, and substituted derivatives [$CH_2=CH$ (acrolein), etc.]

Depending on the polyol used, linear, cyclic, or spiro polyacetals are produced. Mixtures of diols may be used to yield random copolymeric polyacetals.

The condensation reactions are usually carried out in the melt or at the boiling temperature of a solvent in the presence of an acid catalyst such as p-toluenesulfonic acid (PTSA). The solvents usually used are toluene or benzene, so that the water of condensation can be azeotropically removed.

Acetal interchange is also used with the higher aldehydes and for the self-condensation of hydroxyacetals to give polyacetals.

The use of dialdehydes and polyhdroxy compounds yields cyclic acetals, as shown in Eq. (8):

$$\begin{array}{c} HOCH_2 \\ \diagdown \\ CH-R-CH \\ \diagup \\ HOCH_2 \end{array} \begin{array}{c} CH_2OH \\ \diagup \\ \\ \diagdown \\ CH_2OH \end{array} + O=CH-R'-CH=O \xrightarrow{-2H_2O}$$

$$\left[\begin{array}{c} CH_2-O \\ \diagup \diagdown \\ CH CH-R'-CH \\ \diagdown \diagup \\ CH_2-O \end{array} \begin{array}{c} O-CH_2 \\ \diagup \diagdown \\ CH-R \\ \diagdown \diagup \\ O-CH_2 \end{array} \right]_n \quad (8)$$

$R = CH_2$, alkylene, arylene
$R' =$ aliphatic and aromatic groups

15-2. Preparation of a Polyacetal by the Reaction of Benzaldehyde with Bis(2-hydroxyethyl) Sulfide [21]

$$C_6H_5CH=O + HOCH_2CH_2S-CH_2CH_2OH \xrightarrow{-H_2O}$$

$$HOCH_2CH_2 \left[S-CH_2CH_2-O-\underset{\underset{C_6H_5}{|}}{CHO}-CH_2CH_2 \right]_n SCH_2CH_2OH \quad (9)$$

To a resin flask equipped with a mechanical stirrer, thermometer, condenser, Dean & Stark trap, and nitrogen inlet/outlet are added 53.0 gm (0.5 mole) of benzaldehyde, 61.0 gm (0.5 mole) of bis(2-hydroxyethyl) sulfide, 300 ml of toluene, and 0.25 gm of p-toluenesulfonic acid. The mixture is heated to reflux and the water (9.0 ml) of condensation is removed azeotropically over a period of 70 hr. The first 4.5 ml of water (0.25 mole) is obtained in 4 hr and the remaining 0.25 mole is obtained after 3 days of reaction. The polymer is isolated by distillation of the toluene at atmospheric pressure. The product is then heated to 180°C at 1.0 mm Hg for 5 hr. A liquid, nonviscous polymer is isolated with a 69.4 average hydroxyl number, corresponding to a number-average molecular weight of 1610. The infrared spectrum showed absorption at 9.0 μm for ether linkage and a peak at 2.9 μm for the presence of the hydroxyl group.

15-3. Preparation of a Polyacetal by the Reaction of Pentaerythritol with a Mixture of 1,3- and 1,4-Cyclohexane Dialdehydes [22]

$$HOCH_2-\underset{\underset{CH_2OH}{|}}{\overset{\overset{CH_2OH}{|}}{C}}-CH_2OH + O=CH-\left\langle\right\rangle-CH=O \longrightarrow$$

$$\left[-CH\underset{O-CH_2}{\overset{O-CH_2}{\diagup\diagdown}}C\underset{CH_2-O}{\overset{CH_2-O}{\diagup\diagdown}}CH-\left\langle\right\rangle-\right]_n \quad (10)$$

To a flask equipped with mechanical stirrer, condenser, and vacuum connection are added 6.8 gm (0.05 mole) of pentaerythritol and 7.0 gm (0.05 mole) of a mixture of 1,3- and 1,4-cyclohexane dialdehydes. After refluxing for 1 hr, the mass is heated under a nitrogen stream for 1 hr at 110°–180°C. The mass becomes very viscous and then solid, and finally the temperature is raised to 240°C. The temperature is maintained at 240°–250°C for 1 hr, and the pressure reduced to 12 mm Hg. The temperature is now raised to 260°C over a 1.5 hr period, and the mixture is finally cooled to give a brown hard resin of an extremely high softening point (> 250°C).

3. POLY(VINYL ACETALS)

Poly(vinyl alcohol) reacts as a 1,3-diol with aldehydes to form cyclic acetals. The acetalization reaction is sometimes between polymer chains, and the resultant crosslinking reaction renders the polymer insoluble.

In most cases poly(vinyl acetal) resins contain 4–25% hydroxyl groups, on a weight basis, calculated as poly(vinyl alcohol). Poly(vinyl acetals) were first prepared by Hermann and Haehnel [23] in 1924 by a reaction of benzaldehyde with poly(vinyl alcohol). The commercialization of poly(vinyl acetals) began in the 1930s when poly(vinyl alcohol) was reacted in solution with aldehydes (formaldehyde, acetaldehyde, butyraldehyde) in the presence of mineral acids to give products useful for electrical insulation [24, 25].

The five major methods of preparing poly(vinyl acetals) involve the following:

1. Conversion of poly(vinyl acetate) to poly(vinyl alcohol) in acid solution followed by acetalization [26, 27].
2. Reaction of an aqueous poly(vinyl alcohol) solution with the aldehyde until precipitation of the acetal occurs [28].
3. A method similar to method 2 except that a solvent is added for the acetal that is also miscible with water, thereby preventing precipitation.
4. Heterogeneous reaction of film or fiber of poly(vinyl alcohol) with the aldehyde to form the acetal.
5. Poly(vinyl alcohol) is suspended in a suitable nonsolvent, which dissolves the aldehyde and the final product.

The first two methods are more widely used for commercial manufacture of poly(vinyl acetals).

Some of the aldehydes used to acetalize poly(vinyl alcohols) are formaldehyde [29], acetaldehyde [30], propionaldehyde [30], butyraldehyde [30], heptanol [30], palmitic and stearyl aldehydes [31], glyoxal [32], benzaldehyde [23], p-tolualdehyde [33], 2-naphthaldehyde [33], vinyl benzaldehyde [34], and glyoxalic acid [35]. Ketones such as cyclohexanone [36] and methyl ethyl ketone have been used to give ketals [37].

The hydrolysis of various poly(vinyl acetals) in ethanolic 85% phosphoric acid at reflux temperatures has been measured by the percentage of aldehyde liberated in a given length of time (Eq. 11) [38]. The order of resistance to hydrolysis of the polymers examined is poly(vinyl formal) > poly(vinyl propional) > poly(vinyl acetal) > poly(vinyl butyral).

$$+ RCHO + H^+ \quad (11)$$

15-4. General Procedure for the Preparation of Poly(vinyl acetals) [26]

Acetalization (see Note) was carried out in a three-necked resin flask equipped with a mechanical stirrer and condenser. The reaction times vary from 4 to 40 hr at 60°–75°C. At the end of this time water is slowly added to a vigorously stirred solution to precipitate the polymer in a granular form. The heat stability is enhanced by steeping for approximately 4 hr with a 0.02 N potassium hydroxide solution at 50°C. The resin is finally washed with water and dried at 55°–60°C. The hydroxyl content is determined by the acetalation.

NOTE: Acetals of acetaldehyde, propionaldehyde, and n- and isobutyraldehyde are prepared by condensation with PVA in the presence of sulfuric acid in ethanol. Acetals of 2-ethyl butyraldehyde, n-hexaldehyde, n-heptaldehyde, and 2-ethyl hexaldehyde are prepared in a mixture of dioxane and ethanol.

15-5. Preparation of Poly(vinyl butyral) [39]

$$\left[\begin{array}{c} CH_2-CH-CH_2-CH \\ | \quad\quad\quad\quad | \\ OH \quad\quad\quad OH \end{array} \right]_n + nC_3H_7CH=O \longrightarrow$$

$$\left[\begin{array}{c} CH_2 \\ \diagup\diagdown \\ CH_2-CHC \\ | | \\ OO \\ \diagdown\diagup \\ CH \\ | \\ C_3H_7 \end{array} \right] + H_2O \quad (12)$$

To a resin flask equipped with a mechanical stirrer and condenser are added 400 gm (9.0 moles) of poly(vinyl alcohol), 2210 gm of ethanol, 250 gm (3.47 moles of n-butyraldehyde, and 10 gm of sulfuric acid. The reaction mixture is stirred and warmed at 75°–78°C for 2 hr. After this time water is added to precipitate the acetal. The product is washed first with water, then with a warm (80°C), dilute aqueous potassium hydroxide solution to neutralize the residual acid, and finally dried [39–44].

A structural study of poly(vinyl formal) and poly(vinyl butyral) by NMR spectroscopy has shown that the 1,3-dioxane rings of these resins only exhibited the 2,5-twist conformation (racemic tt conformation), even though earlier work had considered the meso *tt* (*cis* chair) form to be favored thermodynamically [40].

REFERENCES

1. J. Read, *J. Chem. Soc.* **101**, 2090 (1912).
2. H. Orth, *Kunstoffe* **41**, 454 (1951).
3. S. M. Cohen and E. Lavin, *J. Appl. Polym. Sci.* **23**, 503 (1962).
4. S. M. Cohen, C. F. Hunt, R. E. Kass, and A. H. Markhart, *J. Appl. Polym. Sci.* **6**, 508 (1962).
5. S. M. Cohen and E. Lavin, U.S. Patent 2,963,464 (1960); British Patent 896,254 (1962).
6. J. W. Hill and W. H. Carothers, *J. Am. Chem. Soc.* **57**, 925 (1935).
7. W. H. Carothers, U.S. Patent 2,071,252 (1937).
8. S. R. Sandler and W. Karo, "Organic Functional Group Preparations," 2nd ed., Vol. 3, Chapter 1, Academic Press, San Diego, 1989; S. R. Sandler and W. Karo, "Polymer Syntheses," 2nd ed., Vol. 1, Chapter 5, Academic Press, San Diego, 1992.
9. H. S. Hill and H. Hibbert, *J. Am. Chem. Soc.* **45**, 3117 and 3124 (1923).
10. R. L. Adelman, U.S. Patent 2,682,532 (1954).
11. N. G. Gaylord, *Encycl. Polym. Technol.* **10**, 319 (1969); *in* "Polyethers" (N. G. Gaylord, ed.), Part I, Vol. XIII, Chapter VII, Wiley (Interscience), New York, 1963; R. W. Lenz, "Organic Chemistry of Synthetic High Polymers," pp. 125–128, Wiley, New York, 1967; H. Gibellow, *Off. Matieres Plast.* **6**, 1258 (1959); D. Sek, *Wiad. Chem.* **26**, 677 (1972).
12. S. Penczek and P. Kubisa, *in* "Ring-Opening Polymerization" (T. Saegusa and E. Goethals, eds.), ACS Symposium Series, No. 59, pp. 60ff, Washington, 1977.
13. R. C. Schulz, K. Albrecht, C. Rentsch, and Q. V. Tran Thi, *in* "Ring-Opening Polymerization" (T. Saegusa and E. Goethals, eds.), ACS Symposium Series, No. 59, pp. 77ff, Washington, 1977.
14. Y. Yamashita and Y. Kawakami, *in* "Ring-Opening Polymerization" (T. Saegusa and E. Goethals, eds.), ACS Symposium Series, No. 59, pp. 99ff, Washington, 1977.
15. R. Szymanski, P. Kubisa, and S. Penczek, *Macromolecules* **1983**, 1000 (1983).
16. T. J. Dolce and J. A. Grates, "Concise Encyclopedia of Polymer Science and Engineering," pp. 4–5, John Wiley, New York, 1990.
17. M. Okada, *Adv. Polym. Sci.* **102**, 1–46 (1992).
18. R. B. Seymour and G. S. Kishenbaum, "High Performance Polymers: Their Origin and Development," pp. 105ff, Elsevier Science Publ. Co., New York, 1986.
19. B. H. Kress, U.S. Patent 2,786,081 (1957).
20. B. H. Kress and E. Abrams, U.S. Patent 2,785,947 (1957).
21. E. Schonfeld, *J. Polym. Sci.* **49**, 277 (1961).
22. L. S. Abbott, D. Faulkner, and C. E. Hollis, U.S. Patent 2,739,972 (1956).
23. W. O. Herrmann and W. Haehnel, German Patent 480,866 (1924).

24. H. Hopff, U.S. Patent 1,955,068 (1934); H. Bauer, J. Heckmaier, R. Reinecke, and E. Bergmeister, German Patent 929,643 (1952).
25. G. O. Morrison, F. W. Skirrow, and K. B. Blackie, U.S. Patent 2,036,092 (1935).
26. A. F. Fitzhugh and R. N. Crozier, *J. Polym. Sci.* **8**, 225 (1952).
27. A. F. Fitzhugh and R. N. Crozier, *J. Polym. Sci.* **9**, 96 (1952).
28. K. Rosenbusch, W. Pense, and F. Winkler, German Patent 1,069,385 (1959).
29. R. D. Dunlop, Fiat Final Report No. 1109 (1947).
30. S. Okamura and T. Motoyama, *Kyogyo Kagaku Zasshi* **55**, 774 (1952).
31. K. Noma and T. Sone, *Kobunshi Kagaku* **4**, 50 (1947).
32. S. Okamura, T. Motoyama, and K. Uno, *Kogyo Kagaku Zasshi* **55**, 776 (1952).
33. E. T. Cline and H. B. Stevenson, U.S. Patent 2,606,803 (1953).
34. E. L. Martin, U.S. Patent 2,929,710 (1954).
35. G. Kranzlein and U. Campert, German Patent 729,774 (1937).
36. G. Kranzlein, A. Voss, and W. Starck, German Patent 661,968 (1930).
37. J. D. Ryan, U.S. Patent 2,425,568 (1947).
38. R. A. Barnes, *J. Polym. Sci.* **27**, 285 (1958).
39. E. Lavin, A. T. Marinaro, and W. R. Richard, U.S. Patent 2,496,480 (1950).
40. P. G. Berger, E. E. Rensem, G. C. Leo, and D. J. David, *Macromolecules* **24**, 2189 (1991).
41. E. Schacht, G. Desmartes, E. Goethals, and T. St. Pierre, *Macromolecules* **16**, 291 (1983).
42. S. Ohmori, S. Ito, and M. Yamamoto, *Macromolecules* **23**, 4047 (1990).
43. J. H. Hopkins and G. H. Wilder, U.S. Patent 2,282,057 (1942).
44. B. C. Bren, J. H. Hopkins, and G. H. Wilder, U.S. Patent 2,282,026 (1942).

16
POLY(VINYL ETHERS)

The first vinyl ether to be polymerized was reported about 120 years ago by Wislicenus [1], who treated ethyl vinyl ether with iodine and obtained a violent reaction giving a resinous material. Later studies [2, 3] on this reaction indicate that I^+ is the active initiator and that carbonium ions are involved.

Reppe [4–7] and co-workers described the polymerization of a wide variety of vinyl ethers by acidic reagents used in small amounts. Some typical acidic reagents or catalysts used by Reppe are $SnCl_4$, $AlCl_4$, BF_3, BF_3 complexes, $FeCl_3$, $ZnCl_2$, $SnCl_4$, H_2SO_4, SO_2, H_3PO_4, and the like. The polymerizations were reported to be violent at room temperature and above and afforded low molecular weight resins. Friedel–Crafts catalysts were used by Favorski and Shostakovskii [8, 9] to polymerize methyl, ethyl, isopropyl, butyl, amyl, and other vinyl ethers.

Eley and Pepper [10] polymerized vinyl n-butyl ether in bulk and in petroleum ether over the temperature range of 20°–94°C using a 2% solution of $SnCl_4$ in petroleum ether as the catalyst. The polymerization kinetics of vinyl 2-ethylhexyl ether was studied similarly [3].

Anionic catalysts fail to initiate polymerization and the mechanism involving Grignard agent catalysts is not clear at this time.

Vinyl ethers are also polymerized by free-radical initiators in bulk, solution, or emulsion to give low molecular weight polymers.

Based on S. R. Sandler and W. Karo, "Polymer Syntheses," 2nd ed., Vol. II, pp. 239ff, by permission of Academic Press, San Diego, 1994.

Vinyl ethers copolymerize well with a wide variety of monomers (such as olefins, haloolefins, alkoxybutadienes, acrylates, acrylonitrile, maleic anhydride, maleates, fumarates, allyl, vinyl pyrrole, and vinyl carbazole) using free-radical ionic initiators or coordination-type catalysts [11].

Vinyl ether polymers are useful in lacquer resins, plasticizers, adhesives, paints, copolymer compositions such as those for fire retardants, marine coatings, anticorrosion agents, thickening agents, and other uses [12, 13].

1. CATIONIC POLYMERIZATION

Vinyl octadecyl ether was the first vinyl ether to be polymerized commercially and boron trifluoride was used as the catalyst [14]. The polymerization was carried out at 90°–95°C, and low molecular weight polymers were obtained with K values of 20–30, m.p. 50°C (white, brittle solid). Other similar products were suggested in patents by I. G. Farbenindustrie, and also the copolymerization with maleic anhydride was mentioned.

Vinyl methyl ether was commercially polymerized in bulk using a 3% solution of boron trifluoride: water (1:2) in dioxane as the initiator [15]. The catalyst was added pointwise at 5°–12°C with cooling, and the polymerization temperature was then allowed to rise to 100°C. The polymer had a balsam-like consistency and was soluble in cold water.

Schildknecht and co-workers [16] were the first to report the preparation of crystalline, isotactic poly(isobutyl vinyl ether).

Methyl vinyl ether yields a crystalline polymer only when methylene chloride is present as a solvent. However, ethyl, isopropyl, and n-butyl vinyl ethers do not yield crystalline polymers. Branched alkyl vinyl ethers, other than isobutyl vinyl ether, and benzyl ether also yield crystalline polymers [17]. The crystallinity of the polymers (isotactic) is similar in soluble and insoluble catalyst systems [18].

Alkyl vinyl ethers can also be copolymerized cationically with such monomers as styrene using $SnCl_4/AlCl_3$ catalyst and in ethyl chloride solution [18–21].

- **CAUTION:** Care should generally be exercised when acidic agents are used to initiate vinyl ether polymerization. Some polymerizations may occur violently, especially in the case of bulk polymerizations. To avoid this situation, low temperatures and diluents or solvents should be used and the reaction carried out on a small scale using adequate precautions behind a shield in a hood.

16-1. Preparation of a Poly(vinyl *n*-butyl ether) Using Stannous Chloride Catalyst [6]

$$n\text{-}C_4H_9\text{-}OCH=CH_2 \xrightarrow{SnCl_2} \left[\begin{array}{c} CH_2-CH- \\ | \\ OC_4H_9 \end{array} \right]_n \quad (1)$$

To 50 gm (0.50 mole) of vinyl *n*-butyl ether is added 0.25 gm of stannous chloride. After a few minutes the temperature begins to rise to 45°–50°C and then more rapidly to 140°C. The product obtained is a yellowish brown, balsam-like mass. Purification by steam distillation leaves 35 gm (70%) of a viscous, liquid, yellow polymerization product.

Similar results are obtained with stannic chloride or aluminum chloride.

16-2. Preparation of Poly(vinyl *n*-butyl ether) Using Boron Trifluoride Etherate Catalyst [22]

$$C_4H_9\text{-}OCH=CH_2 \xrightarrow{BF_3\cdot O(C_2H_5)_2} \left[\begin{array}{c} CH_2-CH- \\ | \\ OC_4H_9 \end{array} \right]_n \quad (2)$$

To a pressure reactor containing 200 gm (2.0 moles) of vinyl *n*-butyl ether are added 600 gm of propane and 750 gm of dry ice. The reactor is cooled to $-80°C$ by dry ice–acetone, and the $BF_3 \cdot O(C_2H_5)_2$ maintained at 25°C is added dropwise until 5 parts have been added. The reaction is complete in 1 hr. The catalyst is deactivated by adding 10 gm of 30% ammonium hydroxide at $-50°C$ followed by 300 gm of propane at $-60°C$. The liquid propane containing the polymer is removed to give 85% (170 gm) of high molecular weight polymer and 1.5% of low molecular weight polymer.

2. FREE-RADICAL POLYMERIZATION

Vinyl ethers polymerize slowly when free-radical initiators are used. The polymers are usually low molecular weight oily substances. The use of heat, light, or radiation gives the same results [23]. Nelson, Banes, and FitzGerald in 1961 reported that 2-10% di-*tert*-butyl peroxide or *tert*-butyl or cumene hydroperoxide gives low molecular weight polymers useful as lubricating oils [24]. Azobis(isobutyronitrile) has been reported to show some ability to effect free-radical polymerization of vinyl ethers [25].

Vinyl ethers copolymerize with methyl methacrylate to give alternating copolymer units as well as homopolymer segments of methyl methacrylate [26]. With other monomers this is also true, and either solution, emulsion, suspension, or bulk polymerization techniques may be used. The pH should be kept at about 8 or above in aqueous systems to prevent hydrolysis of the vinyl ether. The r_2 (alkyl vinyl ether) values are low and approach zero for bulk polymerization systems [27-29] utilizing such monomers as acrylonitrile, butyl maleate, methyl acrylate, methyl methacrylate, styrene, vinyl acetate, vinyl chloride, or vinylidene chloride.

Free-radical copolymerization of alkyl vinyl ethers has been carried out with the following typical monomers: acrylic acid (bulk and emulsion) [30, 31], acrylonitrile (emulsion) [20, 21], acrylic esters (emulsion) [32], methyl methacrylate (bulk) [33], maleic anhydride (solution) [34], vinyl acetate (bulk and emulsion) [21, 35, 36], and vinyl chloride (emulsion) [20, 28, 37]. The properties of these and other copolymers are described in a technical bulletin by General Aniline & Film Corporation [29].

Other monomers that copolymerize with alkyl vinyl ethers are vinyl ketones [38], acrolein diacetate [39], acrylamide [40], alkoxy 1,3-butadienes [41], butadiene [42], chloroprene [43], chlorotrifluoroethylene [44], tri- and tetrafluoroethylene [45], cyclopentadiene [46], dimethylaminoethyl acrylate [47], fluoroacrylates [48], fluoroacrylamides [49], *N*-vinyl carbazole [50, 51], triallyl cyanurate [50, 51], vinyl chloroacetate [52, 53], *N*-vinyl lactans [54], *N*-vinyl succinimide [54], vinylidene cyanide [55, 56], and others. Copolymerization is especially suitable for monomers having electron-withdrawing groups. Solution, emulsion, and suspension techniques can be used. However, in aqueous systems the pH should be buffered at about pH 8 or above to prevent hydrolysis of the vinyl ether to acetaldehyde. Charge-transfer complexes have been suggested to form between vinyl ethers and maleic anhydride, and these participate in the copolymerization [57].

16-3. Preparation of Poly(vinyl ethyl ether) Using Di-*tert*-butyl Peroxide Initiator [17]

$$C_2H_5OCH=CH_2 \longrightarrow \left[\begin{array}{c} CH_2-CH- \\ | \\ OC_2H_5 \end{array}\right]_n \quad (3)$$

To a 1.8-liter stainless steel reactor bomb are added 376 gm (500 ml, 5.22 moles) of vinyl ethyl ether, 400 ml of cyclohexane, and 23.8 gm (0.163 mole) of 6.33 wt.% di-*tert*-butyl peroxide. The reactor bomb is sealed and heated at 159°C for 4 hr with agitation in a rocking apparatus. Then the bomb is cooled and the contents removed. The cyclohexane and other volatiles are removed by stripping at 100°C (overhead temperature) at atmospheric pressure to afford 295.4 gm of product (78%).

16-4. Copolymerization of Maleic Anhydride with Ethyl Vinyl Ether [58]

$$C_2H_5OCH=CH_2 + \underset{O}{\underset{\|}{\overset{O}{\bigvee}}}\underset{O}{\overset{O}{\|}} \xrightarrow[59-61°C]{\text{benzoyl peroxide}} \left(\begin{array}{c} CH-CH_2-CH-CH- \\ | \quad\quad\quad | \quad | \\ OC_2H_5 \quad\quad C \quad C \\ \;\;\;\;\;\;\;\;\;\;\;\;\;\;\;\;\; \overset{\|}{O}\;\;\;\overset{}{O}\;\;\;\overset{\|}{O} \end{array}\right)_n \quad (4)$$

To a 250-ml three-necked flask equipped with a magnetic stirring bar, condenser, thermometer, and gas inlet–outlet is added a benzene solution (toluene can be used in its place and is less toxic) of 10.77 gm (0.1098 mole) of maleic anhydride and 0.04514 gm (1.863 × 10^{-4} mole) of benzoyl peroxide. Dry nitrogen (oxygen-free) is flushed for 45 min through the flask, and then 9.11 gm (0.1214 mole) of ethyl vinyl ether is injected into the flask via a septrum seal on one of the necks. The reaction mixture is stirred and heated for 8 hours at 59°–61°C. The white precipitate is filtered and dissolved in 125 ml of acetone. The product is isolated from the acetone by gradually adding 200 ml of ethyl ether, filtering, and drying under reduced pressure. The copolymer is purified by reprecipitating from an acetone (125 ml) solution by adding it to ethyl ether (150 ml). The precipitate is cut into small pieces, which are immersed in ethyl ether overnight. The polymer is then dried.

These polymers are useful for reacting with alkylamine ($n = 8$, 12) to form amides. Hydrolysis of the remaining anhydride groups gives water soluble polymeric amides.

There has recently been reported a process for preparing high molecular weight copolymers of maleic anhydride and methyl vinyl ether using a free radical initiator in the absence of solvent but using an excess of vinyl ether at about 50–60°C (under pressure) [59].

REFERENCES

1. J. Wislicenus, *Justus Liebigs Ann. Chem.* **192**, 106 (1878).
2. W. Chalmers, *Can. J. Res.* **7**, 113 and 472 (1932); *J. Am. Chem. Soc.* **56**, 912 (1934); D. D. Eley and J. Saunders, *J. Chem. Soc.* p. 4167 (1952); *ibid.* pp. 1668, 1672, and 1677 (1954); D. D. Eley and A. Seabrooke, *ibid.* p. 2226 (1964).
3. D. D. Eley and A. W. Richards, *Trans. Faraday Soc.* **45**, 425 (1949).
4. W. Reppe and O. Schlichting, U.S. Patent 2,104,000 (1937).
5. W. Reppe and O. Schlichting, U.S. Patent 2,104,001 (1937).
6. W. Reppe and O. Schlichting, U.S. Patent 2,104,002 (1937).
7. W. Reppe and E. Kuehn, U.S. Patent 2,098,108 (1937).
8. A. E. Favorski and M. F. Shostakovskii, *J. Gen. Chem.* (*Engl. Transl.*) **13**, 1 428 (1943).
9. M. F. Shostakovskii and I. F. Bogdanov, *J. Gen. Chem. USSR* (*Engl. Transl.*) **15**, 249 (1942); M. F. Shostakovskii, *ibid.* **20**, 609 (1950).
10. D. D. Eley and D. C. Pepper, *Trans. Faraday Soc.* **43**, 112 (1947).
11. C. E. Shildknecht, "Vinyl and Related Polymers," pp. 593–634, Wiley, New York, 1952; N. M. Bikales, *Encycl. Polym. Sci. Technol.* **14**, 511 (1971).
12. S. A. Miller, "Acetylene," Vol. 2, pp. 242–244, Academic Press, New York, 1966.
13. "Alkyl Vinyl Ethers," Technical Bulletin 7543-055, Commercial Development Department, GAF Corporation, New York, 1966.
14. FIAT 856.B105 1602.
15. BIOS 742 and 1292. FIAT 1602.
16. C. E. Schildknecht, A. O. Zoss, and C. McKinley, *Ind. Eng. Chem.* **39**, 180 (1947); C. E. Schildknecht, S. T. Gorss, H. R. Davidson, J. M. Lambert, and A. O. Zoss, *ibid.* **40**, 2104 (1948); C. E. Schildknecht, S. T. Gorss, and A. O. Zoss, *ibid.* **41**, 1998 (1949); C. E. Schildknecht, A. O. Zoss, and F. Grosser, *ibid.* p. 2891.
17. C. E. Schildknecht, *Ind. Eng. Chem.* **50**, 107 (1958).
18. G. Natta, I. Bassi, and P. Corradini, *Makromol. Chem.* **18–19**, 455 (1955); S. Okamura, T. Higashimura, and I. Sakurada, *J. Polym. Sci.* **39**, 507 (1959).

19. H. Fikentscher, German Patent 634,408 (1936); H. G. Hammon, R. A. Clark, and J. W. Uttley, Jr., U.S. Patent 2,994,681 (1961); R. E. Florin, *J. Am. Chem. Soc.* **71**, 1867 (1949).
20. G. A. Richter, Jr. and G. L. Brown, U.S. Patent 2,869,977 (1959).
21. H. Fikentscher, U.S. Patent 2,016,490 (1935).
22. C. E. Schildknecht, U.S. Patent 2,513,820 (1950).
23. S. H. Pinner and R. J. Worrel, *J. Appl. Polym. Sci.* **2**, 122 (1959); Imperial Chemical Industries, Ltd., British Patent 585,179 (1947); E. I. Du Pont de Nemours, British Patent 586,297 (1947).
24. J. F. Nelson, F. W. Banes, and W. P. FitzGerald, U.S. Patent 2,967,203 (1961).
25. M. F. Shostakovskii and A. V. Borgdanova, *Izv. Akad. Nauk SSSR, Otd. Khim. Nauk* p. 919 (1954); p. 387 (1957); N. M. Bortnick and S. Melamed, U.S. Patent 2,734,890 (1956).
26. A. M. Khomutov, *Izv. Akad. Nauk SSSR, Otd. Khim. Nauk* p. 116 (1961).
27. G. Akazome, *Chem. High Polym.* **17**, 449, 452, 578, 482, 558, and 620 (1960); *Chem. High. Polym., Ind. Chem. Sect.* **62**, 1247 (1959); J. Alfrey, J. J. Bohrer, and H. Mark, *Copolymerization*, New York: Wiley (Interscience), 1952.
28. E. C. Chapin, G. E. Ham, and C. L. Mills, *J. Polym. Sci.* **4**, 597 (1949).
29. "Alkyl Vinyl Ethers," Technical Bulletin 7543-055, GAF Corporation, New York, 1966.
30. J. L. Lang, U.S. Patent 2,937,163 (1960).
31. R. R. Dreisbach and J. F. Malloy, U.S. Patent 2,778,812 (1957).
32. Nitto Electrical Industries, Japanese Patent 21,891 (1963); H. Firkentscher and R. Gäth, German Patent 745,424 (1943).
33. P. J. Stedry, U.S. Patent 2,811,501 (1957).
34. M. F. Shostakovskii and A. M. Khomutov, *Bull. Acad. Sci. USSR* p. 931 (1953); F. Grosser, U.S. Patent 2,694,697 (1954); R. S. Towne, U.S. Patent 2,744,098 (1956); J. J. Giammana, U.S. Patent 2,698,316 (1954); E. Knopf and H. Scholz, German Patent 707,321 (1941).
35. A. M. Khomutov and M. F. Shostakovskii, *Izv. Akad. Nauk SSSR, Otd. Khim. Nauk* p. 2017 (1959); A. M. Khomutov, *ibid.* p. 352 (1961); *Polym. Sci. USSR (Engl. Transl.)* **5**, 181 (1964).
36. S. Imada, Japanese Patent 548 (1955).
37. M. F. Shostakovskii, *Zh. Prikl. Khim.* **28**, 1123 (1953).
38. J. M. Wilkinson, Jr. and J. P. Barker, U.S. Patent 2,655,267 (1954).
39. L. M. Minsk and C. C. Unruh, U.S. Patent 2,417,404 (1947).
40. W. Zerweck and W. Kanze, German Patent 948,282 (1956).
41. R. F. Heck, U.S. Patent 3,025,276 (1962); J. Lal, U.S. Patent 3,038,889 (1962).
42. S. N. Ushakov, S. P. Mitsengendler, and V. N. Krasulina, *Izv. Akad. Nauk SSSR, Otd. Khim. Nauk* p. 490 (1957).
43. S. P. Mitsengendler, *Izv. Akad. Nauk SSSR, Otd. Khim. Nauk* p. 1120 (1956).
44. J. J. Robertson, U.S. Patent 2,905,660 (1959).
45. F. Grosser, U.S. Patent 2,547,819 (1951).
46. Mitsubishi Chem. Industrie, British Patent 863,237 (1961).
47. C. S. Scanley, F. H. Siegele, and R. L. Webb, U.S. Patent 3,088,931 (1958).
48. F. W. Knobloch and H. C. Hamlen, P. B. Report 131,998 from *U.S. Gov. Res. Rep.* **31**, 159 (1959).

49. F. W. Knobloch, *J. Polym. Sci.* **25**, 453 (1957).
50. K. Takakura, *Polym. Lett.* p. 565 (1965).
51. D. E. Jefferson, French Patent 1,350,905 (1964).
52. M. F. Shostakovskii, A. M. Khomutov, and P. Alimov, *Izv. Akad. Nauk SSSR, Ser. Khim.* p. 1839 (1963).
53. F. P. Sidel'kovskaya, M. F. Shostakovskii, F. Ibinginov, and M. A. Askarov, *Vysokomol. Soedin.* **6**, 1585 (1964).
54. J. Furukawa, T. Tsuruta, H. Fukutani, and N. Yamamoto, *Kogyo Kagaku Zasshi* **60**, 1085 (1957).
55. B. F. Goodrich, British Patent 756,839 (1956).
56. F. F. Miller and H. Gilbert, German Patent 953,660 (1956).
57. S. Iwatsuki and Y. Yamashita, *J. Polym. Sci., Polym. Chem. Ed.* **5**, 1733 (1967); K. Ohara, N. Sugiyama, M. Nakamura, and I. Kaneko, *Asahi Garasu Kenkyn Hokoko* **41**(1), 51 (1991); *Chem. Abstr.* **116**(4), 129738q (1972).
58. Y. Chang and C. L. McCormick, *ACS Polymer Reprints* **31**(2), Aug. 1990, p. 462. Paper presented at the Washington D.C. meeting—Division of Polymer Chemistry.
59. Z. Pehlah, I. Potencalk, W. Kopp, and E. Urmann, U.S. Patent 5,047,490 (1991).

17
POLY(N-VINYLPYRROLIDONE)

Monomeric N-vinylpyrrolidone (NVP) was one of the materials that resulted from the extensive developments of "Reppe" chemistry pursued during the second world war in Germany. The polymer was found to be water soluble and of a low order of toxicity. In some respects it resembled albumen. At the I. G. Farbenindustrie operations in Elberfeld and in Ludwigshafen, considerable quantities of the resin were produced. In the form of a 2.5% saline dispersion, it was used under the trade name Periston as a blood substitute that could be used regardless of blood type. It has been reported that up to 50% of the total blood content of the body could be replaced with Periston in battlefield transfusions in the German army [1]. Reference [1], which was based on interviews at German installations by U.S. personnel as the war was winding down, seemed to have been unable to get information on where or how the testing on men and women was accomplished.

Poly(N-vinylpyrrolidone) (PVP) has not been used as a blood substitute for some time. However, since the polymers and copolymers are thought to have a low order of toxicity, applications in many alternative fields have been developed. They are used in cosmetics, toiletries, and a variety of pharmaceuticals, including the manufacture of tablets and microencapsulated materials. Other applications are in such diverse fields as in adhesives, the textile and dyeing industries, suspending agents, protective colloids, pigment dispersants, leveling agents, desalination membranes, surfactants,

---

Based on S. R. Sandler and W. Karo, "Polymer Syntheses," 2nd ed., Vol. II, pp. 261ff, by permission of Academic Press, San Diego, 1994.

flexible contact lenses, and fiberglass sizing. Since the polymer forms complexes with many materials, such as phenolic compounds, crosslinked copolymers are used in beverage clarification. A PVP complex with iodine is used as a household antiseptic [2].

The literature on poly(N-vinylpyrrolidone) is voluminous. References [2-20] represent a selection of the reviews available. Linke's article [18] is of particular interest for its discussion of PVP and its copolymers in hairsprays and some of the problems associated with packaging in aerosol spray cans.

The polymerization of N-vinylpyrrolidone has been initiated by many of the systems conventional for the formation of high polymers of vinyl compounds. The notable exception to conventional initiations is that dibenzoyl peroxide is not a satisfactory initiator for this monomer. Oxidative side reactions may interfere with the formation of a high molecular weight product. On the other hand, hydrogen peroxide (particularly in the presence of ammonia), hydroperoxides, 2,2'-azobis(isobutyronitrile), persulfates, sodium peroxide, sodium sulfite, and the like have been used as polymerization initiators. Ultraviolet radiation, electron beams, γ-radiation, heat (in the presence of oxygen), titanium tetrachloride, boron trifluoride etherate, halides of mercury, antimony, and bismuth, and $Al(C_2H_5)_{1.5}Cl_{1.5}$ have been used to form PVP. Grafting of N-vinylpyrrolidone to various substrates, particularly to textile fibers, has been accomplished.

Bulk polymerization procedures are known. Suspension polymerization in aqueous media loaded with high concentrations of electrolyte are possible. Work has also been carried out on solid-state polymerizations. However, the most important methods of polymerization are solution processes, particularly using water as the solvent.

1. BULK POLYMERIZATION PROCEDURES

17-1. Bulk Polymerization with AIBN [21]

In a suitable glass ampoule are placed 5 gm (45 mmoles) of freshly distilled N-vinylpyrrolidone and 0.01 gm (0.062 mole) of 2,2'-azobis(isobutyronitrile). The ampoule is flushed with nitrogen and sealed. After shaking the reaction to ensure complete dissolution of the initiator, a protective sleeve is placed around the ampoule, and the assembly is heated in an oil bath at 60° ± 1°C for 72 hr. The ampoule is then removed from the bath, cooled, and cautiously opened. The product is dissolved in ethanol and precipitated from solution with ligroin to yield 3.4 gm (68% of theoretical yield).

Bretenbach and Schmidt [22] found that without initiator present, the bulk polymerization of N-vinylpyrrolidone proceeded readily at elevated temperatures (Table I).

Hydrophilic contact lens materials have been prepared by graft polymerizing 2-hydroxyethyl methacrylate onto poly(N-vinylpyrrolidone) using *tert*-butyl perbenzoate as an initiator [23]. A more complex crosslinked contact lens material has been prepared by polymerizing a composition of PVP, N-vinylpyrrolidone, 2-hydroxyethyl methacrylate, and ethylene dimethacrylate with benzoyl peroxide [24].

Because poly(vinylpyrrolidone) is a water-soluble polymer, crosslinked copolymers of this material imbibe water and swell. This fact has been applied to the preparation of contact lenses which are soft and flexible on contact with water or aqueous fluids not only by grafting appropriate monomers to preformed PVP but also by direct copolymerization of N-vinylpyrrolidone with other comonomers. For example, a crosslinked hydrogel was prepared by terpolymerizing N-vinylpyrrolidone, methyl acrylate, and tetraethylene glycol dimethacrylate in the presence of AIBN [25].

We had occasion to extend the concept to the development of materials for potential use in intraocular lenses [26]. Appropriately shaped dry resin lenses were to be surgically implanted. Upon

TABLE I
BULK POLYMERIZATION OF UNINITIATED N-VINYLPYRROLIDONE [22]

Reaction temperature (°C)	Initial rate of polymerization (wt % of polymer formed per hr)	Average degree of polymerization
140	0.026	440
160	0.18	300
180	0.60	250

exposure to the fluids in the eye, the polymer was to swell and assume the correct optical shape for proper vision. This concept required that the polymer should take up at least 40% of its weight as water to produce a swollen polymer that had a refractive index of no less than 1.40 to permit the proper optical corrections to be applied to the lens. Assuming that refractive indices of mixtures or solutions are additive, this requires a polymer with a refractive index of about 1.5 so that 40% of water with η_D 1.33 will result in a hydrogel in the desired refractive index range. Procedure 17-2 is an example taken from a recent patent that bulk polymerizes a composition of five monomers—some water soluble, some with high refractive indices, and one, a high refractive index crosslinking agent—to give a dry and swellable polymer with a refractive index of 1.557. The mention of this patent is only for reference. It is not given here to permit its practice in any form without permission of the patent holders.

17-2. Bulk Polymerization of a PVP-Based Hydrogel [26]

A number of microscope slides were dipped into a solution of lecithin in hexane and then air-dried. From these slides a number of polymerization cells were assembled. Each cell consisted of two slides separated by an elastomeric spacer held together with spring clamps. Thin Neoprene tubing may be used as the spacer. The lecithin coating on the inside of each cell acted as a "parting agent."

A solution of 5.6 gm of N-vinylpyrrolidone, 2.4 gm of acrylamide, 0.80 gm of 2-hydroxypropyl acrylate, 0.80 gm of 2-hydroxyethyl methacrylate, and 0.4 gm of dibromoneopentylglycol dimethacrylate was treated with 0.025 gm of 2,2'-azobisisobutyronitrile (AIBN). The resultant solution was placed in the polymerization cells. The tops of the cells were closed with more of the spacer-elastomer. The cells were then placed in an oven and heated at 60°C for 18–24 hr. The cells were then cooled. The polymer was removed from the cells. The refractive index at 20°C, taken on an Abbé refractometer by adhering the polymer to the measuring prism with 1-bromo-naphthalene and making the measurement with reflected light, was 1.557.

2. AQUEOUS SOLUTION PROCEDURES

For commercial production of PVP, aqueous solution processes are most important. Product isolation may be somewhat difficult in the laboratory since facilities for spray drying or film casting are usually not available. However, techniques such as precipitation methods may be applicable. If the product is of sufficiently high molecular weight and the electrolyte concentration is sufficiently high, poly(N-vinylpyrrolidone) may separate from the solvent spontaneously.

17-3. Generalized Procedure for Polymerization with Ammonia–Hydrogen Peroxide [3]

A review article by Fikentscher and Herrle [3] discusses the solution polymerization of this monomer in considerable detail. This material is summarized here so that readers may develop their own reaction procedures on the basis of the information. The reaction conditions based on this work are also reliable for the preparation of copolymers of N-vinylpyrrolidone and vinyl acetate [27].

Al-Alawi [28] used an aqueous solution procedure similar to that detailed below to prepare several PVPs of differing molecular weights. It is interesting to note that he used these polymers for the template polymerization of acrylic acid. Evidently acrylic acid itself forms a complex with PVP either prior to or during this process. Prepolymerized poly(acrylic acid) itself also forms complexes with PVP. In general, these types of complexes are thought to result from hydrogen bonding between the carboxyllic hydrogen and the carbonyls of the PVP [29].

It is postulated that in water, N-vinylpyrrolidone forms a monohydrate. If hydrogen peroxide is added, the medium becomes acidic —presumably because of the formation of acetic acid from the acetaldehyde that is generated from the monomer. For this reason the reaction medium is usually buffered to maintain the pH between 8 and 7. Ammonia or amines, aside from acting as buffering agents, also have an activating effect.

In the Fikentscher and Herrle treatment, the rate of polymerization is given in units of kilograms of polymer formed per cubic meter of solution per hour. The rate is determined by plotting the amount of polymer formed (in kg/m^3) against time (in hr) (cf. Fig. 1). A tangent is drawn at the point where the straight line portion of the curve begins after curving up from the abscissa. The tangent is extended to the time axis. The trigonometric tangent of the angle θ (Fig. 1) is considered to be the rate of polymerization ($kg/m^3/hr$). The intercept on the time axis is the induction period (min). To be noted is that the period from time 0 to the first perceptible evidence of polymer formation is given no special term.

The percentages of initiators or activators are always given as based on the monomer. Ammonia is given in terms of the weight of gaseous ammonia used, although it is actually added as aqueous ammonium hydroxide solution. The quantity of hydrogen peroxide is not precisely defined but is believed to be in terms of the weight of 30% hydrogen peroxide solution added. The initial pH is always 8 (unless otherwise noted) and is never allowed to drop below 7.

The effect of ammonia on the polymerization characteristics is given in Table II.

When amines are substituted for ammonia in equivalent amounts, the induction period is increased and the rate of polymerization is decreased, but the K value is essentially unchanged. Substituting equivalent quantities of sodium hydroxide or sodium bicarbonate for ammonia did not lead to perceptible polymerization even after 4 hr.

The initial pH of the system affects the induction period and the rate of polymerization but does not have a significant effect on the

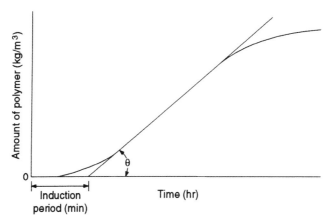

Figure 1 Polymerization of N-vinylpyrrolidone. Tan θ, rate of polymerization in $kg/m^3/hr$.

TABLE II
EFFECT OF AMMONIA ON THE POLYMERIZATION OF N-VINYLPYRROLIDONE[a]

Ammonia added (%)	Induction period (min)	Rate of polymerization (kg/m³/hr)	K value
0[b]	—	No polymerization even after 4 hr	—
0.04[c]	180	200	63
0.1	5	250	53
0.4	0	500	56
1.6	0	850	62

[a] Polymerization conditions: 30% of monomer in distilled water; initial pH = 8 (unless otherwise indicated); 0.5% of 30% H_2O_2; temperature, 50°C [1].
[b] At pH = 7 initially.
[c] Up to 0.5% of sodium bicarbonate had been added to maintain the pH at 7.

K value. Thus under reaction conditions similar to those given in Table II, with 0.1% of ammonia, the induction period above a pH of 9 is 5 min or less, the rate of polymerization is above 200 kg/m³/hr, but the K value remains at 55.

A patent [30] states that batch polymerizations of N-vinylpyrrolidone in aqueous solution with hydrogen peroxide often give rise to gel formation. This difficulty can be overcome by replacing at least part of the water with such substances as isopropyl alcohol, thioglycolic acid, dimethylformamide, ethanolamine, methyl ethyl ketone, trichloroacetic acid, or 2-mercaptoethanol.

3. CATIONIC POLYMERIZATION

Procedure 17-4 represents an early cationic polymerization experiment with N-vinylpyrrolidone.

17-4. Copolymerization N-Vinylpyrrolidone and Methyl Methacrylate with Triethylboron [31]

With suitable safety precautions for the handling of triethylboron, a small three-necked flask, under nitrogen, is charged with 5.0 gm (0.05 mole) of methyl methacrylate and 4.4 gm (0.04 mole) of N-vinylpyrrolidone. With vigorous stirring, the solution is cooled to 0°C. Then, while the reaction temperature is maintained between 0°

and 2°C, 0.53 ml of triethylboron, 0.22 ml of cumene hydroperoxide, and 0.28 ml of pyridine are added. The reaction mixture is stirred for 4 hr at a temperature between 0° and 2°C and then is added cautiously to an excess of methanol. The product is filtered off, dissolved in benzene, and reprecipitated with methanol. After filtration, the product is dried overnight under reduced pressure at 60°C. Yield: 8.7 gm (93%); specific viscosity of a 1% solution in benzene at 25°C is 3.152.

4. COMPLEX FORMATION

The preparation of PVP in aqueous solution and the swelling of copolymers with water was discussed above. These polymers are highly hydrated as they are formed. Such hydrates are reasonably easily dehydrated. Therefore it may be a semantic matter whether such hydrates should be considered to be "complexes." We should note that in our own work, we have frequently found that on swelling NVP copolymers with water, the mass-increase did not always result in corresponding volume changes. This indicated to us that water, at least in some cases, did not simply act to separate strands of the copolymers, as we would have expected in a simple hydration process.

The complex-formation of PVP with poly(acrylic acid) has been mentioned above. In general, mixing the two polymers in aqueous solution led to products that differed from template-polymerized acrylic acid only that some, but not all, of the poly(acrylic acid) was grafted onto the poly(vinylpyrrolidone) [28, 29].

Complexes have been formed by hydrogen bonds or by ion-dipole interactions between a copolymer of methacrylic acid and methacrylamide with a copolymer of acrylic acid and acrylamide. If there are unreacted units left over in such complexes, these, in turn, may undergo further complexation with PVP [32].

The complex formation of PVP with partially hydrolyzed poly(acrylamide) is dependent both on the degree of hydrolysis and on the pH. At low degrees of hydrolysis, the formation is detectable though weak and independent of pH. At high degrees of hydrolysis, the interaction is strong below a pH of 6 [33].

The complex formation of PVP with sodium dodecyl sulfate (SDS) should be of interest not only in its own right but also in connection with the use of mixtures of PVP and SDS or other surfactants in emulsion or suspension polymerizations [34–36].

As sodium dodecyl sulfate is added to a dilute solution of PVP, initially there is no significant interaction. Suddenly SDS molecules associate with the polymer (the "premicelle" stage). This occurs below the critical micelle concentration (cmc) of SDS in water. At high levels of SDS free micelles of SDS coexist in equilibrium with the premicelles.

In effect there are three regions of PVP/SDS interaction. The effect is substantially independent of the molecular weight of the PVP used. However, if we consider a 0.1% solution of PVP with molecular weight 40,000, the first region (the premicelle region) extends from 0 to approximately 10^{-3} moles of SDS (mol. wt. = 287). The second region, which includes the cmc, extends to approx. 4×10^{-3} moles of SDS. The third region comprises SDS concentrations greater than that. The cmc for SDS is given as 8×10^{-8} moles [37].

REFERENCES

1. J. M. DeBell, W. C. Goggin, and W. E. Gloor, "German Plastics Practice," pp. 1 and 154, DeBell & Richardson, Springfield, Massachusetts, 1946.
2. D. H. Lorenz, *Encycl. Poly. Sci. Technol.* **14**, 239 (1971).
3. H. F. Fikentscher and K. Herrle, *Mod. Plast.* **23**(3), November, 157 (1945).
4. W. Reppe, "Polyvinylpyrrolidone," Verlag Chemie, Weinheim, 1954.
5. E. S. Barabas, in "Concise Encyclopedia of Polymer Science and Engineering," pp. 1236ff, Wiley (Interscience), New York, 1990.
6. W. Reppe, "Acetylene Chemistry," P. B. Rep. 18852-S. Charles A. Meyer and Co., Inc., New York, 1949.
7. C. E. Schildknecht, "Vinyl and Related Polymers," p. 662, Wiley, New York, 1952.
8. I. Greenfield, *Ind. Chem.* **32**, 11 (1956).
9. J. Remond, *Rev. Prod. Chim.* **59**, 127 and 260 (1956).
10. E. Ferraris, *Mater. Plast.* (*Milan*) **25**, 208 (1959).
11. W. W. Myddleton, *Manuf. Chem.* **34**(7), 316 (1963).
12. "Polyvinylpyrrolidone," Technical Bulletins 7543-113, 7543-031, 7543-029, and 7543-066, GAF Corporation, New York.
13. F. A. Wagner, *Kirk-Othmer Encycl. Chem. Technol.* **17**, 391 (1968).
14. R. H. Kaplan, Thesis, Cornell University, Ithaca, New York, 1969; University Microfilms, Ann Arbor, Michigan, Order No. 70-5801, *Diss. Abstr. Int. B* **31**, 162 (1970).
15. F. P. Sidel'kovskaya, "Chemistry of *N*-Vinylpyrrolidone and Its Polymers," Nauka, Moscow, 1970; *Chem. Abstr.* **74**, 100, 438v (1971).

16. J. P. Schroeder and D. C. Schroeder, in "Vinyl and Diene Monomers" (E. C. Leonard, ed.), Part 3, pp. 1362ff, Wiley, New York, 1971.
17. H. Warson, *Polym., Paint Color J.* **161**, 637 and 643 (1972).
18. W. Linke, *Chemtech.* **4**, 288 (1974).
19. "V-Pyrol, N-Vinyl-2-Pyrrolidone," Technical Bulletin 9653-011. GAF Corporation, New York.
20. M. M. Flannery, ed., "PVP, Polyvinylpyrrolidone, An Annotated Bibliography to 1950," Commercial Development Department, GAF Corporation, New York, 1951.
21. F. P. Sidel'kovskaya, M. A. Askarov, and F. Ibragimov, *Vysokomol. Soedin.* **6**, 1810 (1964); *Polym. Sci. USSR (Engl. Transl.)* **6**, 2005 (1964); *Chem. Abstr.* **62**, 6563c (1965).
22. J. W. Bretenbach and A. Schmidt, *Monatsh. Chem.* **83**, 833 (1952).
23. D. G. Ewell, U.S. Patent 3,647,736 (1972); *Chem. Abstr.* **76**, 154737k (1972).
24. M. Seiderman, U.S. Patent 3,639,524 (1972); *Chem. Abstr.* **76**, 141762f (1972).
25. R. Steckler, U.S. Patent 3,532,679 (1970).
26. S. B. Siepser, B. D. Halpern, and W. Karo, U.S. Patent 5,147,394 (1992).
27. Author's laboratory (WK).
28. S. S. Al-Alawi, *Macromolecules* **24**, 4206 (1991).
29. S. Al-Alawi and N. A. Saeed, *Macromolecules* **23**, 4474 (1990).
30. J. F. Volks and T. G. Traylor, U.S. Patent 2,982,762 (1961).
31. J. Bond and P. I. Lee, *J. Polym. Sci., Polym. Chem. Ed.* **9**, 1777 (1971).
32. S. K. Chatterjee, A. M. Khan, and S. Ghosh, *Angew. Makromol. Chem.* **200**, 1 (1992).
33. C. Maltesh, P. Somasundaran, Pradip, R. A. Kulkarni, and S. Gundiah, *Macromolecules* **24**, 5775 (1991).
34. N. J. Turro, B. H. Baretz, and P.-L. Kuo, *Macromolecules* **17**, 1321 (1984).
35. E. V. Anufrieva, T. N. Nekrasova, V. B. Lushchik, Yu. A. Fedotov, Yu. E. Kirsh, and M. G. Krakovyak, *Visokomol. Soedin., Ser. B* **344**, 31 (1992).
36. A. J. Paine, W. Luymes, and J. McNulty, *Macromolecules* **23**, 3104 (1990).
37. J. D. Song, R. Ryoo, and M. S. Jhon, *Macromolecules* **24**, 1727 (1991).

18
SILICONE RESINS (POLYORGANOSILOXANES) OR SILICONES

This section describes the topic of silicone resins or polyorganosiloxanes.

$$\left[\begin{array}{c} R \\ | \\ -Si-O- \\ | \\ R \end{array} \right]_n$$

The hydrolysis of organochlorosilanes or organoacetoxysilanes by water can be accelerated by small amounts of acids or alkalis. Preferred acid catalysts are those that are easily removed by washing with water such as HCl, HCOOH, $(COOH)_2$, Cl_3CCOOH, and CH_3COOH [1]:

$$X-\underset{\underset{R}{|}}{\overset{\overset{R}{|}}{Si}}-X + nH_2O \longrightarrow \left[\begin{array}{c} R \\ | \\ -Si-O- \\ | \\ R \end{array} \right]_n + 2nHX \quad (1)$$

Silicone resins differ from the fluids or elastomers in that they are highly crosslinked and contain a high proportion of silicon atoms with one or no organic substituent groups. A typical starting reaction solution may contain CH_3SiCl_3, $(CH_3)_2SiCl_2$, $C_6H_5-SiCl_3$, $(C_6H_5)_2SiCl_2$, and sometimes also $SiCl_4$ dissolved in toluene. The use of various starting materials affects the properties of the final resin in the following ways: (1) Trifunctional siloxy groups give harder, less organic-soluble polymers; (2) difunctional siloxy groups give softer, more organic-soluble polymers; and (3) monophenylsiloxanes give more organic-soluble and thermally stable polymers than the methylsiloxanes.

The toluene solution of the chlorosilanes is hydrolyzed, and the aqueous hydrochloric acid is separated, washed, and then heated in the presence of a mild condensation catalyst to adjust the resin to the proper viscosity and cure time. Fillers can be added prior to the removal of the solvent and isolation of the final resin. A resin can also be prepared from phenyltrichlorosilane alone to give a tough, infusible resin that can be cast from solvents to give a clear film [2, 3].

The final cure step of the silicone resin solution involves either removal of solvent or addition of a metal soap (tin acetate) or amine catalyst prior to heat curing with subsequent solvent evaporation.

In most coatings, the silicone resin is copolymerized with another resin to form silicone/alkyd, silicone/acrylic, or silicone/polyester enamels. The silicone/alkyds are used for air-drying coating applications, whereas the silicone/polyester resin is for heat-cured applications [4].

18-1. Silicone Resins from Phenyltrichlorosilane [2, 5]

$$C_6H_5SiCl_3 \xrightarrow{H_2O} C_6H_5Si(OH)_3 \longrightarrow \text{[crosslinked phenylsiloxane network]} \quad (2)$$

A mixture of the hydrolyzate of phenyltrichlorosilane (300 gm), 0.1% potassium hydroxide, and an equal weight of toluene is refluxed for 16 hr while water is removed in a Dean & Stark trap. The reaction mixture is cooled and filtered, and the product precipitated into ligroin or methanol to give approximately 99.9% condensation of the silanol analyzing for $(C_6H_5-SiO_{3/2})_x$. The polymer obtained had the following properties: $[\eta] = 0.12$ dl/gm in benzene, $\overline{M}_n = 14{,}000$, $\overline{M}_w = 26{,}000$. Similar polymers were prepared from phenyltriethoxysilane, base catalyzed by hydrolysis [5]. See Ref. [2] for more details.

18-2. Silicone Resins by Cyclohydrolysis of Dimethyldichlorosilane and Methyltrichlorosilane [6]

$$(CH_3)_2SiCl_2 + CH_3SiCl_3 \xrightarrow{H_2O} \begin{bmatrix} \begin{array}{c} CH_3 \quad CH_3 \quad CH_3 \\ | \quad\quad | \quad\quad | \\ -Si-O-Si-O-Si-O- \\ | \quad\quad\quad\quad | \\ CH_3 \quad O \quad CH_3 \\ \quad\quad | \\ \quad\quad CH_3 \quad CH_3 \\ \quad\quad | \quad\quad | \\ -O-Si-O-Si-O-Si-O- \\ | \quad\quad | \quad\quad | \\ CH_3 \quad CH_3 \quad O \\ \quad\quad\quad\quad\quad | \end{array} \end{bmatrix}_n \quad (3)$$

A mixture of 4.5 gm (0.035 mole) of dimethyldichlorosilane and 1.95 gm (0.013 mole) of methyltrichlorosilane in 50 ml of ether is hydrolyzed by pouring into 100 gm of cracked ice. The ether is evaporated and the residue heated in air to give a glossy, infusible, and insoluble solid with a C:Si ratio of 1.3. The resin gives silicon when heated at 300°–400°C.

REFERENCES

1. R. R. McGregor and E. L. Warrick, German Patent 881,404 (1942).
2. J. F. Brown, Jr., L. H. Vogt, Jr., A. Katchman, J. W. Eustace, K. M. Kiser, and K. W. Krantz, *J. Am. Chem. Soc.* **82**, 6194 (1960).
3. R. N. Meals and F. M. Lewis, "Silicones," Van Nostrand-Reinhold, Princeton, New Jersey, 1959; F. J. Modic, *Adhes. Age* p. 5 (1962).
4. M. Smart, J. Lutz, and T. Sasano, Silicone resins, *in* "SRI-Chemical Economics Handbook," pp. 580, 1950A, 1988.
5. J. F. Brown, Jr., L. H. Vogt, Jr., and P. I. Prescott, *J. Am. Chem. Soc.* **86**, 1120 (1964).
6. E. G. Rochow and W. F. Gilliam, *J. Am. Chem. Soc.* **63**, 798 (1941).

19
OLEFIN-SULFUR DIOXIDE COPOLYMERS

It was recognized early on that amorphous products were formed by the reaction of olefins and sulfur dioxide [1-4]. The elucidation of the products formed was first reported by Marvel [6] and Staudinger [7, 8] in the early 1930s.

Marvel reported that propylene and cyclohexene react with sulfur dioxide to form alternating copolymers of olefin and sulfur dioxide in a head-to-tail arrangement [6, 9]. Staudinger reported that 1,3-butadiene reacts with sulfur dioxide to form a cyclic sulfone and an amorphous linear polysulfone [7, 8, 10] (see Eqs. 1 and 2).

$$\diagdown C = C \diagup + SO_2 \longrightarrow \left[SO_2 - \overset{|}{\underset{|}{C}} - \overset{|}{\underset{|}{C}} \right]_n \quad (1)$$

$$CH_2 = CH - CH = CH_2 + SO_2 \longrightarrow (SO_2 - CH_2 - CH = CH - CH_2)_n + \underset{\text{Sulfolene} \\ \text{m.p. 65°C}}{\boxed{SO_2}}$$

(2)

Fitch [11] at Phillips Petroleum Co. was issued a patent in 1936 that was followed by many others on the formation of polymeric sulfones from the reaction of olefins and sulfur dioxide. A patent to Frey and Snow of Phillips Petroleum Co. also describes some of the prior literature [12].

The olefin-sulfur dioxide polysulfone compositions are usually light-colored, thermoplastic, amorphous products that are moldable and extrudable. The polymers are not very resistant to alkaline solutions.

The only olefin-sulfur dioxide product sold commercially is sulfolene by Phillips Petroleum Co. The linear polymers remain to be commercially produced. It is interesting to note that a polysulfone, i.e., poly(phenylene sulfone), has been sold commercially since 1966 [13].

For more detailed discussions and historical background several earlier reviews are worth consulting [14, 15].

Sulfur dioxide does not homopolymerize, but on reaction with olefins it yields copolymers. Terminal olefins react more readily than those with an internal double bond. The presence of various substituents affects the rate of polymerization. Conjugated dienes copolymerize with sulfur dioxide to give linear polymers containing residual double bonds.

The copolymerization reaction is free radical in nature and is catalyzed by such initiators as peroxides [16], oxygen, azo compounds [17, 18], and light [15, 19]. Styrene dissolved in liquid sulfur dioxide and catalyzed by stannic chloride (cationic catalyst) gives only polystyrene whereas the use of azobisisobutyronitrile gives poly(styrene sulfone).

Sulfur dioxide is a gas at ordinary temperatures and has the following physical properties:

m.p.	$-75.46°C$
b.p.	$-10.02°C$ (density at $-10°C = 1.46$ gm/ml)
vapor pressure	28.5 cm Hg at $-30°C$
	53.1 cm Hg at $-20°C$
	115 cm Hg at $0°C$
	171 cm Hg at $10°C$
	2456 cm Hg at $20°C$

Reactions with sulfur dioxide are carried out in pressure vessels constructed of glass or stainless steel. Caution should be used and all reactions should be done under a hood and behind a barrier for protection.

Copolymerization is usually carried out in solution or emulsion [20–22], but not in the vapor phase [23]. The reaction in the gaseous phase usually gives sulfinic acids. A solid-state copolymerization has been reported for vinyl acetate with sulfur dioxide initiated by radiation [24].

$$\text{Liquid phase:} \quad SO_2 + \!\!\!\diagdown\!\!\!\!C\!\!=\!\!C\!\!\!\diagup\!\!\!\longrightarrow \diagdown\!\!\!\!\cdot C\!-\!C\!\!-\!SO_2 \xrightarrow{\diagdown\!\!C=C\!\!\diagup / SO_2} \left[\!\!\diagdown\!\!\!\!C\!-\!C\!\!\!-\!SO_2\!\!\right]_n \quad (3)$$

$$\text{Gas phase:} \quad SO_2 + \!\!\!\diagdown\!\!\!\!C\!\!=\!\!C\!\!\!\diagup\!\!\!\longrightarrow \diagdown\!\!\!\!C\!\!=\!\!C\!\!\!\diagup_{SO_2H} \quad (4)$$

A ceiling temperature [25] (temperature at which rate of propagation and depropagation are equal) is independent of the kind of initiator, but depends on the given olefin. The polymerization is not always accelerated by heat and in fact in some cases slows as it approaches the ceiling temperature.

The ceiling temperature for the copolymerization of sulfur dioxide with various vinyl compounds is said to follow the following relationship [26]:

$$T_c = \Delta H / S$$

where ΔH is the heat change and S is the entropy change.

Some characteristic infrared absorption frequencies in the 7.5–9.5 micron range are shown for several olefin polysulfones in Fig. 1. Two strong absorption bands, one at 7.7 microns and the other at 8.75–9.0 microns, are characteristic stretching vibrations of the sulfone group as shown in Fig. 1.

19-1. Preparation of Octene–Sulfur Dioxide Copolymers [27]

$$C_6H_{13}\!-\!CH\!=\!CH_2 + SO_2 \longrightarrow \left[\!\!\begin{array}{c} C_6H_{13} \\ | \\ C\!-\!CH_2\!-\!SO_2 \\ | \\ H \end{array}\!\!\right]_n \quad (5)$$

To a cooled, glass pressure vessel is added 40 gm (0.62 mole) of liquid sulfur dioxide, 30 gm (0.27 mole) of octene, and 0.3 gm *tert*-butyl hydroperoxide. The vessel is sealed and the contents agitated and warmed to room temperature. After 50 hr, the con-

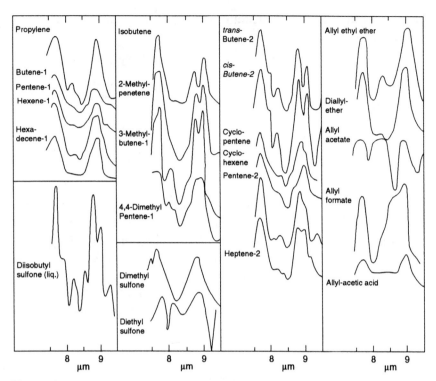

Figure 1 Infrared absorption spectra of some olefin polysulfones and some aliphatic sulfones. Absorption increasing upward. [Reprinted from R. E. Cook, F. S. Dainton, and K. J. Ivin, *J. Polymer Sci.* **26**, 351 (1957). Copyright 1957 by the *Journal of Polymer Science*. Reprinted by permission of the copyright owner.]

tainer is vented and the polymer placed in a vacuum oven to strip off excess sulfur dioxide and unreacted olefin. The polymer yield is 45.6 gm (95%).

19-2. Preparation of Propylene–Sulfur Dioxide Copolymers [28]

$$CH_2=CH-CH_3 + SO_2 \longrightarrow \left[\begin{array}{c} CH_3 \\ | \\ CH-CH_2-SO_2 \end{array} \right]_n \quad (6)$$

To a 1-liter, stainless steel pressure rocker bomb cooled with dry ice–acetone is added 33 ml (0.41 mole) of liquid propylene followed by 33 ml (0.75 mole) of liquid sulfur dioxide. Then 4 ml of paraldehyde is added and the bomb sealed. The bottle is rocked without cooling for approx 2–3 hr. The bomb is cooled and opened to afford a solid. The solid is ground finely in a mortar, washed with two 50 ml portions of ethyl ether and dried to give 40 gm (83%) of the white amorphous polymer, PMT approx 250°–270°C (decomp). The polymer is insoluble in chloroform, carbon tetrachloride, dioxane, ether, ethyl acetate, benzene, acetone, and acetyl chloride. The polymer dissolves in concentrated sulfuric acid without change.

The use of 1 ml of ascaridole and 7 ml of absolute ethanol can also initiate the polymerization which requires 24 hr at 25°–30°C [9]. A recent review of this area is worth consulting [29].

REFERENCES

1. W. Solomina, *J. Russ. Phys.-Chem. Soc.* **30**, 826 (1898).
2. Badische Aniline und Soda-Fabrik Akt.-Ges. German Patent 236,386 (1910).
3. C. Harries, *Justus Liebigs Ann. Chem.* **383**, 166 (1911); I. Ostromuistenskii, *J. Russ. Phys. Chem. Soc.* **47**, 1983 (1915); W. F. Seyer and L. Hodnelt, *J. Am. Chem. Soc.* **58**, 996 (1936); H. J. Backer and J. Strating, *Recl. Trav. Chim. Pays-Bas* **54**, 170 (1935); G. W. Fenton and C. K. Ingold, *J. Chem. Soc.* p. 2338 (1929).
4. F. E. Matthews and H. M. Elder, British Patent 11,635 (1914).
5. W. F. Seyer and E. G. King, *J. Am. Chem. Soc.* **55**, 3140 (1933).
6. D. S. Frederick, H. D. Cogan, and C. S. Marvel, *J. Am. Chem. Soc.* **56**, 1815 (1934).
7. H. Staudinger and B. Ritzentheler, *Ber. Dtsch. Chem. Ges. B* **68**, 455 (1935).
8. H. Staudinger, German Patent 506,839 (1929); French Patent 698,857 (1930).
9. C. S. Marvel and E. D. Weil, *J. Am. Chem. Soc.* **76**, 61 (1954).
10. S. F. Birch and D. T. McAllen, *J. Chem. Soc.* p. 2556 (1951).
11. L. H. Fitch, Jr., U.S. Patent 2,045,592 (1936); Phillips Petroleum Co., French Patent 808,580 (1937).
12. F. E. Frey, R. D. Snow, and L. H. Fitch, Jr., U.S. Patent 2,198,936 (1940).
13. E. W. Krummel, *Mod. Plast.* **43**, 335 (1966); Anonymous, *ibid.* **42**, 87 (1965); *Chem. Eng. News* **43**, 28 (1965).
14. O. J. Grumitt and A. E. Ardis, *J. Chem. Educ.* **23**, 73 (1946); C. S. Marvel, in "Organic Chemistry" (H. Gilman, ed.), 2nd ed., pp. 765–767, Wiley, New York, 1943; C. S. Suter, "The Organic Chemistry of Sulfur," pp. 757–761, Wiley, New York, 1944; N. Tokura, *Encycl. Polym. Sci. Technol.* **9**, 460 (1968); E. Wellisch, E. Gripstein, and O. J. Sweeting, *J. Appl. Polym. Sci.* **8**, 1623 (1964); W. W. Crouch and J. E. Wicklatz, *Ind. Eng. Chem.* **47**, 160 (1955); F. S. Dainton and K. J. Ivin, *Q. Rev. Chem. Soc.* **12**, 61 (1958).
15. R. D. Snow and F. E. Frey, *Ind. Eng. Chem.* **30**, 176 (1938).
16. C. S. Marvel, L. F. Audrieth, and W. H. Sharkey, *J. Am. Chem. Soc.* **64**, 1229 (1942); Ruhröl G.m.b.H., German Patent 885,778 (1953); *Chem. Abstr.* **51**, 1656 (1957).

17. M. A. Naylor, Jr. and A. W. Anderson, *J. Am. Chem. Soc.* **76**, 3962 (1954); F. S. Dainton and K. J. Ivin, *Discuss. Faraday Soc.* **14**, 199 (1953).
18. C. S. Marvel and F. J. Glavis, *J. Am. Chem. Soc.* **60**, 2622 (1938).
19. R. D. Snow and F. E. Frey, *J. Am. Chem. Soc.* **65**, 2417 (1943); F. S. Dainton and K. J. Ivin, *Proc. R. Soc. London, Ser. A* **212**, 96 and 207 (1952).
20. W. W. Crouch and J. E. Wicklatz, *Ind. Eng. Chem.* **47**, 160 (1955).
21. W. W. Crouch, U.S. Patents 2,593,414 and 2,602,782 (1952).
22. W. W. Crouch and J. F. Howe, U.S. Patent 2,556,799 (1951).
23. F. S. Dainton and K. J. Ivin, *Trans. Faraday Soc.* **46**, 374 and 382 (1950).
24. Z. Kuri and T. Yoshumura, *J. Poly. Sci. part B* **1**, 107 (1963).
25. R. E. Cook, F. S. Dainton, and K. H. Ivin, *J. Polym. Sci.* **26**, 351 (1957).
26. R. E. Cook, F. S. Dainton, and K. H. Ivin, *J. Polym. Sci.* **26**, 351 (1957).
27. D. N. Gray, U.S. Patent 3,728,185 (1973).
28. M. Hunt and C. S. Marvel, *J. Am. Chem. Soc.* **57**, 1691 (1935).
29. E. D. Weil and S. R. Sandler, in "Kirk-Othmer Encyclopedia of Chemical Technology," 4th ed. vol. 23, pp. 267–340, Wiley, New York, 1997.

20
SULFIDE POLYMERS

The sulfide polymers covered in this chapter are those based on the polysulfides derived from the reaction of dihalides and sodium di- or polysulfides; polyphenylene sulfides generated from aryl dihalides and sodium sulfide; polyalkylene sulfides generated from ethylene episulfide or dimercaptan–olefin or diolefin–diketone reactions, all of which are outlined in Scheme 1.

The sulfide polymers are used because of their resistance to oils and solvents, and they are used as specialty plastics [poly(phenyl sulfides)] or as gaskets, hoses, printing rolls, etc. (polysulfides).

Polymers based on ethylene sulfide (thiirane, ethylene episulfide) have not developed into commercial products yet.

Polysulfides are used in synthetic rubber compositions [1], epoxy resin modifiers [2], coatings [3], adhesives [4], sealants [5], and many other products as described in recent *Chemical Abstract* issues.

For additional background material up to about 1960 one should consult available reviews [6–9]. This chapter covers the literature up to 1976.

1. CONDENSATION REACTIONS

The raw materials for the preparation of the polysulfides are sodium polysulfide and dichloroalkane. Sodium polysulfide is prepared by reacting sulfur with aqueous caustic to give sodium polysulfides of various ranks (value of x in the empirical formula Na_xS_x) (see Eq. 1).

$$6NaOH + (2_x + 2)S \rightarrow 2Na_2S_x + Na_2S_2O_3 + 3H_2O \tag{1}$$

SCHEME 1
PREPARATION OF SULFIDE POLYMERS

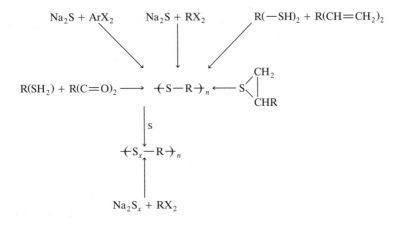

Heating for 1 hr gives a rank of 4 and heating for 6 hr gives a rank of 2. The sodium thiosulfate is not removed since it does not interfere in the subsequent reaction with organic dihalides. The latter reaction is rapid and is complete in 1 hr at 120°F.

The organic dihalides most often used are those that are commercially available such as methylene chloride, ethylene dichloride, propylene dichloride, glycerol dichlorohydrin, dichloroethyl ether, dichloroethylformal, and triglycol dichloride. The most reactive chloro-compounds are the primary ones and these are the most commonly used.

Dichloroethylformal can be prepared by reacting ethylene chlorohydrin with formaldehyde and the water of reaction is removed either by azeotropic distillation with solvent (ethylene dichloride) or by vacuum distillation (see Eq. 2).

$$2Cl-CH_2CH_2OH + H_2C=O \xrightarrow{-H_2O} (Cl-CH_2CH_2O)_2CH_2 \qquad (2)$$

The polymerization can be carried out with or without dispersing agents, but the latter is a more commonly used process in which magnesium hydroxide from magnesium chloride and sodium hydroxide is used in the presence of surfactants and/or wetting agents.

2. OXIDATION REACTIONS

Martin and Patrick [10] reported that the ethylene disulfide polymer prepared from ethylene dichloride and sodium disulfide is practically identical to the polymer prepared by the sodium hypobromate oxidation of alkaline solutions of ethylene mercaptan. The resulting polymer can be converted to the tetrasulfide [11] by combination with 2 gm atoms of sulfur. In addition the tetrasulfide can be reconverted to the disulfide polymer by treatments with sodium hydroxide (Eq. 3).

$$\begin{array}{c} HS-CH_2-CH_2-SH \xrightarrow{[O]} \\ Cl-CH_2-CH_2Cl + Na_2S_2 \longrightarrow {+\!\!(CH_2CH_2-SS)_n} \\ \Big\downarrow 2S \quad\quad \Big\uparrow {NaOH \atop -2S} \\ Cl-CH_2CH_2-Cl + Na_2S_4 \longrightarrow \left(CH_2-CH_2-\underset{\underset{S}{\|}}{S}-\underset{\underset{S}{\|}}{S}\right)_n \end{array} \quad (3)$$

These polysulfides are usually powdery or rubber-like.

20-1. Preparation of a Polydisulfide by the Oxidation of Hexamethylenedithiol with Bromine [12]*

$$HS-(CH_2)_6-SH + Br_2 \to H\text{+}S-(CH_2)_6-S\text{+}_n H \quad (4)$$

Five milliliters of hexamethylenedithiol was added to 50 ml of distilled water containing a few drops of an antifoam and 1.5 gm lauric acid in a 4 oz. polymerization bottle. Four grams of potassium hydroxide was added and the mixture shaken until solution was complete. Then, air was bubbled through the mixture for 4–10 days. [Also possible is oxidation via bromine (approx. 6 gm of bromine was added and the bottle was capped and shaken on a mechanical shaker for 3 hr).] Some polymer formed almost immediately and separated as precoagulum. At the end of the shaking period alum coagulant [12, 13] was added to break the emulsion and the polymer was collected on a filter. The precipitate contained a great deal of

---

*Reprinted in part from C. S. Marvel and L. E. Olson, *J. Am. Chem. Soc.* **79**, 3089 (1957). Copyright 1957 by the American Chemical Society. Reprinted by permission of the copyright owner.

inorganic material. By extracting with chloroform and precipitating in methanol, about 50% of the theoretical yield of polymer was obtained as a white powder with an inherent viscosity of 0.226 (chloroform); softening point, 57°C.

See Ref. [12] for details on preparation of the alum coagulant.

More recently, the oxidation of mono- and dimercaptans to disulfides has been reported [14, 15]. The dimercaptans afford polydisulfides whereas the monomercaptans give disulfides [15].

3. POLY(ARYLENE SULFIDES)

The search for a thermally stable thermoplastic polymer led to the recent developments in poly(phenylene sulfides). The latter polymers are analogous to the poly(phenylene ethers) described in an earlier volume of this series [16].

Poly(p-phenylene sulfide) was first reported in 1897 by Genvresse [17] who reported an insoluble resin prepared by the reaction of benzene with sulfur in the presence of aluminum chloride. A variety of other procedures were reported to yield similar resins. Macallum [18] in 1948 reported a novel procedure that yielded an improved resin. Lenz and co-workers [19–21] modified the procedure and Edmonds and Hill [22] of the Phillips Petroleum Co. developed a commercially successful process. The material is now marketed under the trade name Ryton [23]. The crystallinity of the polymer has recently been reported [24–26].

Earlier suggested syntheses for poly(arylene sulfides) are shown in Scheme 2. Most of these syntheses involve either electrophilic or thermal reactions. Macallum [18, 35] reported a more convenient poly(phenylene sulfide) syntheses by the reaction of p-dichlorobenzene in a dry state using a mixture of sulfur and sodium carbonate at 300°–340°C (Eq. 5). Macallum reported that if sodium sulfide were used in place of sodium carbonate and sulfur then a small amount of sulfur was still required to catalyze the reaction (Eq. 6).

$$Cl-\underset{}{\bigcirc}-Cl + Na_2CO_3 + S \longrightarrow \left[\underset{}{\bigcirc}-S\right]_n \quad (5)$$

SCHEME 2
Earlier Preparations of Poly(arylene Sulfides) [17, 27–34]

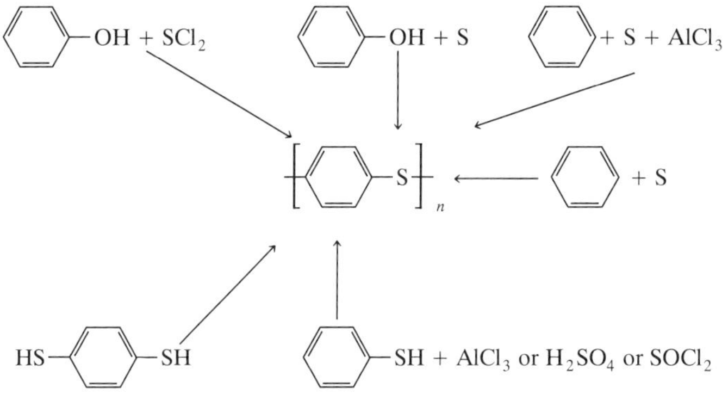

⟨C₆H₄⟩—OH + SCl₂   ⟨C₆H₄⟩—OH + S   ⟨C₆H₄⟩ + S + AlCl₃

[⟨C₆H₄⟩—S]ₙ ← ⟨C₆H₄⟩ + S

HS—⟨C₆H₄⟩—SH   ⟨C₆H₄⟩—SH + AlCl₃ or H₂SO₄ or SOCl₂

The reaction of sulfur and sodium carbonate gives the nascent metal sulfide as earlier described by Pearson and Robinson [36].

$$3Na_2CO_3 + (2n + 2)S \longrightarrow 2Na_2S_n + Na_2S_2O_3 + 3CO_2 \quad (6)$$

The poly(phenylene sulfides) are light-colored, cream to canary yellow, in the solid state having good thermal stability. The addition of sulfur acts as a plasticizer to give a poly(phenylene sulfide) with rubber-like properties. The infrared spectra of phenylene sulfide polymers are chosen in Fig. 1.

Edmonds and Hill of Phillips Petroleum Co. reported more recently that poly(phenylene sulfides) can be prepared by the reaction of *p*-dihalobenzene with sodium sulfide in *N*-methylpyrrolidone or dimethylformamide [37, 38]. Copper metal or cuprous chloride was also shown by Edmonds and Hill to give improved yields [37]. Table I lists examples of various poly(phenylene sulfides) and the conditions used for their preparation. Phillips Petroleum Co. is now producing poly(phenylene sulfides) under the same Ryton (trademark of Phillips Petroleum Co.) [23].

20-2. Preparation of Poly(phenylene sulfide) by the Reaction of *p*-Dichlorobenzene and Sodium Sulfide in *N*-Methyl-2-pyrrolidone [37, 38]

In a stainless steel bomb located behind a safety shield in a hood is added 60 gm $Na_2S \cdot 9H_2O$ in 100 ml *N*-methyl-2-pyrrolidone. The mixture is heated to 160°–190°C while flushing with nitrogen, to remove water of hydration from the sodium sulfide. To the resulting

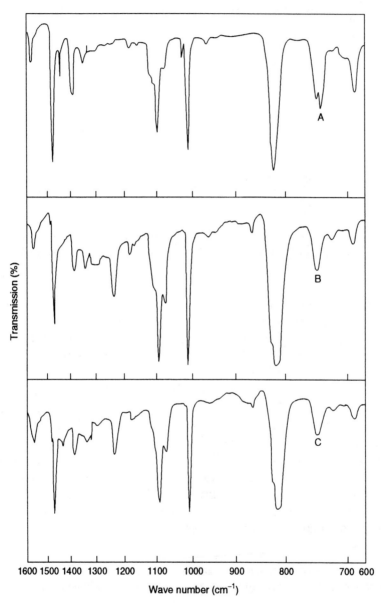

Figure 1 Infrared spectra of phenylene sulfide polymers: (A) linear polymer, (B) Macallum homopolymer, (C) Macallum copolymer (Nujol mulls). [Reprinted from R. W. Lenz and Handlovits, *J. Polymer Sci.* **43**, 167 (1960). Copyright 1960 by the *Journal of Polymer Science*. Reprinted by permission of the copyright owner.]

TABLE I
PREPARATION OF POLYPHENYLSULFIDES USING PROCEDURE 20-2[a]

$Na_2S \cdot 9H_2O$ (gm)	p-Dichlorobenzene (or other, as specified) (gm)	Additive (gm)	Volume of solvent (ml)	Reaction Time (°C)	Reaction Time (hr)	Polymer (gm)	Yield (%)	m.p. (°C)	Low shear visc at 303°C Poises
73.4	45	—	200[b]	260–265	41.5	26	79	275	8.8
73.4	45	—	200[b]	260	91	20	60.5	248.5–252	—
240.2	147	—	1,000[b]	250	17	96.3	—	275	—
240.2	147	—	1,200[c]	250	16	59.9	—	240	—
240.2	147	—	1,000[b]	2150	17	91.0	84.2	282.5–285	40
240.2	147	—	1,000[b]	250	17	90.7	84.0	286–291	37.3
240.2	147	—	1,000[b]	250	17	91.2	84.5	284–287	37.8
240.2	147	CuCl(2)	1,000[b]	250	17	92.8	—	282.5–286.5	18.5
240.2	147	CuCl(4)	1,000[b]	250	17	92.6	—	285.5–286.5	52.4
240.2	147	Cu(tubing)	1,000[b]	250	17	98.8	—	276–280	56.3
240.2	147	CuCl(8)	1,000[b]	250	17	89.6	—	282.5–288.5	12.4
240.2	73.5 p-dichlorobenzene,		100[b]	250	17	100.3	89.5	< 100	—
73.2	100 bis(p-bromophenyl)ether		400[b]	250	17	47.1	—	195–200.5	—

[a] Data from J. T. Edmonds, Jr. and H. W. Hill, Jr., U.S. Patent 3,354,129 (1967).
[b] N-Methylpyrrolidone.
[c] DMF.

solution is added 36.7 gm (0.25 mole) of *p*-dichlorobenzene and the resulting mixture is sealed and heated at 231°C for 44 hr. Then the mixture is heated for 20 hr at 225°C followed by heating for 24 hr at 260°C. The bomb is cooled, opened cautiously, and then the polymer is removed, washed with water, acetone, and dried. The polymer melting point is 275°–285°C and it can be molded at 290°C to a hard film. Some typical reaction conditions, polymer yield, and the effect of copper or copper salts are shown in Table I.

REFERENCES

1. A. Y. Coran, *Rubber Dev.* **18**(1), 4 (1965).
2. J. S. King and R. McDowall, U.S. Patent 3,014,007 (1958).
3. B. Zdralek, H. Stock, and J. Schmidt, German Patent 1,164,008 (1964); *Chem. Abstr.* **60P**, 16072e (1964).
4. R. Hayat and C. Moussafir, Fr. Addn. 81,902 (1963); *Chem. Abstr.* **59P**, 15465 (1963).
5. A. A. Duell, *Paint, Oil Colour J.* **141**, 1558 (1962); *Chem. Abstr.* **58A**, 1640g (1963).
6. M. B. Berenbaum, in "Polyethers" (N. G. Gaylord, ed.), Part 3, Chapter 13, Wiley (Interscience), New York, 1962.
7. J. R. Panek, in "Polyethers" (N. G. Gaylord, ed.), Part 3, Chapter 14, Wiley (Interscience), New York, 1962.
8. M. B. Berenbaum, *Encycl. Polym. Sci. Technol.* **11**, 425 (1969).
9. E. M. Fettes and J. S. Jorczak, in "Polymer Processes" (C. E. Schildknecht, ed.), pp. 425–498, Wiley (Interscience), New York, 1956.
10. S. M. Martin, Jr. and J. C. Patrick, *Ind. Eng. Chem.* **28**(10), 144 (1936).
11. J. R. Katz, *Trans. Faraday Soc.* **32**, 77 (1936).
12. C. S. Marvel and L. E. Olson, *J. Am. Chem. Soc.* **79**, 3089 (1957).
13. C. S. Marvel, V. C. Menikheim, H. K. Inskip, W. K. Taft, and B. G. Labbé, *J. Polym. Sci.* **10**, 39 (1953).
14. T. J. Wallace, U.S. Patent 3,376,313 (1968).
15. J. V. Karabinos and C. N. Yiannios, U.S. Patent 3,513,088 (1970).
16. S. R. Sandler and W. Karo, "Polymer Syntheses," Vol. 1, pp. 239–247, Academic Press, New York, 1974.
17. P. Genvresse, *Bull. Soc. Chem. Fr.* **17**, 599 (1897).
18. A. D. Macallum, *J. Org. Chem.* **13**, 154 (1948).
19. R. W. Lenz, C. E. Handlovits, and H. A. Smith, *J. Polym. Sci.* **58**, 351 (1962).
20. R. W. Lenz and W. K. Carrington, *J. Polym. Sci.* **48**, 333 (1959).
21. R. W. Lenz and C. E. Handlovits, *J. Polym. Sci.* **43**, 167 (1960).
22. J. T. Edmonds and H. W. Hill, Jr., U.S. Patent 3,354,129 (1967).
23. Technical Bulletin Ryton PPS TSM-266, Phillips Chem. Co., Bartlesville, Oklahoma, 1976.
24. D. G. Brady, *J. Appl. Polym. Sci.* **20**, 2541 (1976).
25. B. J. Tabor, E. P. Magre, and J. Born, *Eur. Polym. J.* **7**, 1127 (1971).

26. S. Tsumawaki and C. C. Price, *J. Polym. Sci., Part A* **2**, 1511 (1964).
27. R. W. Lenz, *in* "Polyethers" (N. G. Gaylord, ed.), pp. 30–41, Wiley (Interscience), New York, 1962.
28. C. Friedel and J. M. Crafts, *Ann. Chim. Phys.* [6] **14**, 433 (1888); S. Onufrowicz, *Ber. Dtsch. Chem. Ges.* **23**, 3369 (1890).
29. I. Boeseken, *Recl. Trav. Chim. Pays-Bas* **24**, 6 (1905).
30. G. Dougherty and P. D. Hammond, *J. Am. Chem. Soc.* **57**, 117 (1935); M. P. Genvresse, *Bull. Soc. Chim. Fr.* [3] **15**, 1038 (1896).
31. H. B. Glass and E. E. Reid, *J. Am. Chem. Soc.* **51**, 3428 (1929).
32. J. J. B. Dress, *Recl. Trav. Chim. Pays-Bas* **28**, 136 (1909); H. S. Tasker and H. O. Jones, *J. Chem. Soc.* **95**, 1910 (1909); T. P. Hilditch, *ibid.* **97**, 2579 (1910).
33. R. Leuckart, *J. Prakt. Chem.* [2] **41**, 206 (1890).
34. C. Ellis, "The Chemistry of Synthetic Resins," Vol. II, pp. 1183 and 1189, Van Nostrand-Reinhold, Princeton, New York, 1935.
35. A. D. Macallum, U.S. Patents 2,513,188 (1950); 2,538,941 (1951).
36. T. G. Pearson and P. L. Robinson, *J. Chem. Soc.* pp. 413, 1304, and 1473 (1931).
37. J. T. Edmonds, Jr. and H. W. Hill, Jr., U.S. Patent 3,354,129 (1967).
38. J. T. Edmonds, Jr. and H. W. Hill, Jr., U.S. Patent 3,524,835 (1970).

21
POLYMERIZATION OF MONO- AND DIISOCYANATES

The major homopolymerizations reactions of isocyanates are shown in Scheme 1. These reactions are discussed in this chapter. A good review of the chain structure, polymerization, and conformation of polyisocyanates should be consulted for further details on the 1-nylon-type polymers [1].

In most cases, especially in the presence of catalysts, it is possible that more than one of the isocyanate reactions shown in Scheme 1 takes place simultaneously.

1. HOMOPOLYMERIZATION OF MONOISOCYANATES TO 1-NYLON

Shashona [2, 3] reported that the anionic-catalyzed polymerization of monoisocyanates at $-20°$ to $-100°C$ gave linear high molecular weight polymers (see Table I and Eq. 1). The reaction takes place with both aliphatic and aromatic isocyanates and the polymers are classified as 1-Nylons:

$$R-N=C=O \longrightarrow \left(\begin{matrix} O \\ \| \\ N-C \\ | \\ R \end{matrix} \right)_n \quad (1)$$

Suitable catalysts are shown in Table II, where the solvent is generally DMF. Some typical polymer preparations are also shown in Table II. Table II illustrates the use of different solvents and

SCHEME 1

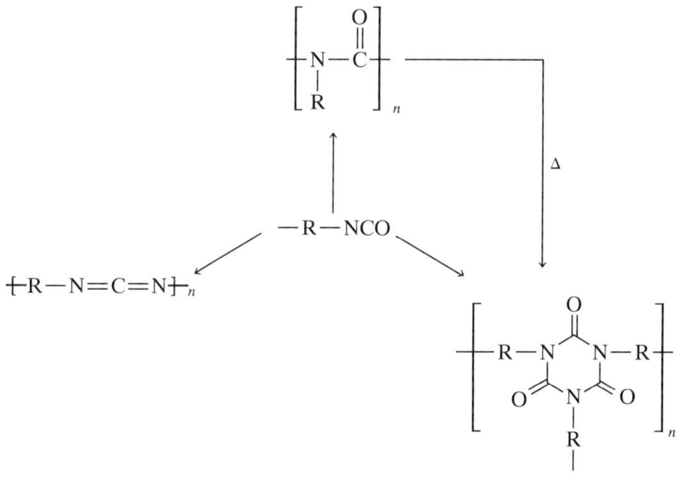

$$\left[\begin{array}{c} \text{N}-\overset{\overset{\displaystyle O}{\|}}{\text{C}} \\ | \\ \text{R} \end{array} \right]_n$$

$\{\text{R}-\text{N}=\text{C}=\text{N}\}_n \longleftarrow -\text{R}-\text{NCO} \longrightarrow$

$$\left[\begin{array}{c} \text{R}-\text{N} \underset{\underset{\displaystyle R}{|}}{\overset{\overset{\displaystyle O}{\|}}{\underset{\displaystyle \|}{\text{C}}}} \text{N} \underset{\displaystyle O}{\overset{\displaystyle C}{\|}} \text{N}-\text{R} \end{array} \right]_n$$

catalysts to effect the polymerization, which is thought to take place as follows (Eq. 2):

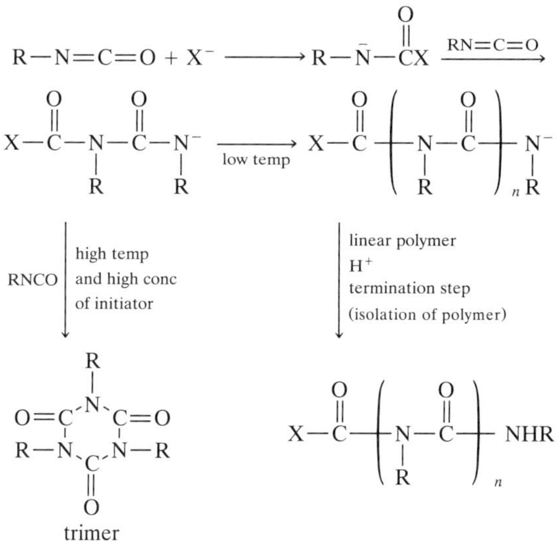

$$R-N=C=O + X^- \longrightarrow R-\bar{N}-\overset{\overset{\displaystyle O}{\|}}{C}X \xrightarrow{RN=C=O}$$

$$X-\overset{\overset{\displaystyle O}{\|}}{C}-\underset{\underset{\displaystyle R}{|}}{N}-\overset{\overset{\displaystyle O}{\|}}{C}-\underset{\underset{\displaystyle R}{|}}{N^-} \xrightarrow{\text{low temp}} X-\overset{\overset{\displaystyle O}{\|}}{C}\left(\underset{\underset{\displaystyle R}{|}}{N}-\overset{\overset{\displaystyle O}{\|}}{C}\right)_n \underset{\underset{\displaystyle R}{|}}{N^-} \quad (2)$$

RNCO | high temp and high conc of initiator

linear polymer
H^+
termination step
(isolation of polymer)

$$\begin{array}{c} \text{R} \\ | \\ O=C\overset{N}{\underset{\underset{\displaystyle \|}{\underset{\displaystyle O}{C}}}{\diagdown}} C=O \\ R-N \diagup \quad \diagdown N-R \end{array}$$
trimer

$$X-\overset{\overset{\displaystyle O}{\|}}{C}\left(\underset{\underset{\displaystyle R}{|}}{N}-\overset{\overset{\displaystyle O}{\|}}{C}\right)_n \text{NHR}$$

21-1. General Method for the Preparation of 1-Nylon [2]

To a 250-ml, three-necked flask (thoroughly dried by flame) equipped with a stirrer, two side-adapters (one for calcium chloride

TABLE I
1-NYLON POLYMERS AND THEIR PROPERTIES[a]

$$\left[\begin{array}{c} -\text{N}-\text{C}- \\ | \quad \| \\ \text{R} \quad \text{O} \end{array} \right]_n$$

R	Softening[b] temp (°C)	Melting[c] or decompn temp (°C)	Soluble in
Ethyl		250	H_2SO_4, CF_3CO_2H
n-Propyl	180	250	H_2SO_4, CF_3CO_2H
n-Butyl	180	209	Aromatic and chlorinated hydrocarbons
Isobutyl	173	210–220	
n-Amyl	145	209	Same
n-Hexyl	120	195	Same
n-Heptyl	100	180	Same
n-Undecyl	45	155	Same
n-Octadecyl	40	94	Same
Allyl	180	260–290	H_2SO_4, CF_3CO_2H
9-n-Decenyl	75	180–240	Aromatic and chlorinated hydrocarbons
Benzyl	—	250	H_2SO_4, CF_3CO_2H
Phenyl	—	197	H_2SO_4
m-Methylphenyl	—	200	N,N-Dimethylformamide
p-Methoxyphenyl	—	212–214	N,N-Dimethylformamide

[a] Reprinted from V. E. Shashona, W. Sweeney and R. F. Tietz, *J. Am. Chem. Soc.* **82**, 866 (1960). Copyright 1960 by the American Chemical Society. Reprinted by permission of the copyright owners.

[b] Temperature at which the polymer first became plastic without sticking on a metal temperature gradient bar.

[c] Temperature at which the polymer first left a clear molten trail on a metal temperature gradient bar.

and a low-temperature thermometer, the other with a nitrogen bubbler and rubber, hypodermic syringe cap) is added 30 ml DMF, and the contents cooled to $-58°C$. A hypodermic needle is used to pierce the rubber cap and deliver the catalyst (see Table II) dropwise over a 2 to 3 min period while the reaction mixture is stirred vigorously. The polymer precipitates rapidly, and after 15 min of stirring at $-58°C$, 50 ml methanol is added. The polymer is filtered and washed with 300 ml methanol, dried at 40°C under vacuum, and isolated. A group of typical preparations are described in Table II, using DMF as a solvent. Other solvents were less effective.

TABLE II
POLYMERIZATION OF THE ISOCYANATES[a]

Expt	Isocyanate	Amount of RNCO	DMF (ml)	Initiator (ml)	Temp (°C)	Polymer yield (%)	η_{inh}	Solvent[b]
1	Ethyl	25 gm	25[c]	10[e]	−100	41	0.3[h]	TFA
2	Ethyl	9 gm	25	2[f]	−50	48	—	—
3	n-Propyl	10 ml	30	2[g]	−58	70	0.02[i]	H_2SO_4
4	Isopropyl	10 ml	30	2[g]	−58	None		
5	n-Butyl	10 ml	30	1[g]	−55	75	15.7[i]	Benzene
6	Isobutyl	10 gm	41	7[f]	−50	18	0.2[i]	H_2SO_4
7	n-Amyl	10 ml	20	5[f]	−50	67	8.6[i]	Benzene
8	n-Hexyl	8 ml	30	4[g]	−58	85	2.9[i]	Benzene
9	n-Heptyl	8.6 gm	41	5[f]	−51	59	2.4[i]	Benzene
10	n-Undecyl	8.5 gm	50	1[f]	−30	70	4.8[i]	Benzene
11	n-Octadecyl	10 ml	30	1[g]	+20	2	0.4[h]	Benzene
12	n-Octanoyl	10 ml	30	3[g]	−40	None	—	—
13	Allyl	8 gm	30	2[g]	−40	10	0.6[i]	H_2SO_4
14	9-n-Decenyl	10 ml	30	2[g]	−58	75	5.2[i]	Benzene
15	Carbethoxymethyl	10 ml	30[d]	3[g]	−70	70	0.4[h]	H_2SO_4
16	Cyclohexyl	10 ml	30	3[g]	−40	None	—	—
17	Phenyl	25 gm	25	1[g]	−40	28	—	—
18	m-Tolyl	10 ml	30	8[g]	−58	60	0.3	DMF
19	p-Tolyl	5 ml	50	1[g]	−50	25	1.1[h]	—[j]
20	m-Chlorophenyl	10 ml	30	8[g]	−20	None	—	—
21	p-Chlorophenyl	10 ml	30	4[g]	−30	None	—	—
22	o-Methoxyphenyl	10 gm	30	4[g]	−40	None	—	—
23	p-Methoxyphenyl	35 ml	100	12[g]	−58	35	0.7[h]	Benzene
24	Benzyl	12.5 ml	30	1[g]	−40	50	0.04[h]	H_2SO_4
25	Benzoyl	10 ml	30	3[g]	−40	None	—	—
26	α-Naphthyl	10 ml	30	3[g]	−40	None	—	—
27	α-(3-Isocyanatoethyl)-phenyl	10 ml	45	2.25[g]	−58	11	—	—
28	2,4-Toluene di-	10 ml	30[c]	3[g]	−100	30	—	—
29	Trimethylsilyl	10 ml	30	4[g]	−40	None	—	—

[a] Reprinted from V. E. Shashona, W. Sweeney, and R. F. Tietz, *J. Am. Chem. Soc.* **82**, 866 (1960). Copyright 1960 by the American Chemical Society. Reprinted by permission of the copyright owner.
[b] DMF = N,N-dimethylformamide, DMA = N,N-dimethylacetamide, TFA = trifluoroacetic acid.
[c] Triethylamine used as solvent.
[d] DMF/DMA (70/30 by vol) used as solvent.
[e] Sodium in DMF.
[f] Sodium benzophenone ketyl.
[g] NaCN in DMF.
[h] At 0.5% concn.
[i] At 0.1% concn.
[j] sym-Tetrachloroethane–phenol (66:100 by wt).

Heating 1-Nylon at elevated temperatures converts it to trimers (isocyanurates).

Diisocyanates on homopolymerization produce crosslinked polymers. The reaction is usually carried out in DMF with sodium cyanide initiator at low temperature [4].

Cyclopolymerization can also occur with 1,2- [4, 5] and 1,3-diisocyanates [4] as well as 1,2,3-triisocyanates [5, 6] to give cyclic ureas (Eq. 3).

$$\text{NCO} \quad \text{NCO} \longrightarrow \left[\begin{array}{c} \overset{O}{\underset{\|}{C}} \quad \overset{O}{\underset{\|}{C}} \\ -N \quad N- \end{array} \right]_n \quad (3)$$

2. POLYMERIZATION OF DIISOCYANATES TO GIVE POLYCARBODIIMIDES

Campbell and co-workers [7–9] reported that certain phospholenes and phosphalene oxides [9] are useful in converting diisocyanates to polycarbodiimides of high molecular weight (Eq. 4). The polymers were reported to be molded into tough, clear, nylon-like films having good tensile strength and electrical properties. The carbodiimide reaction has also been used in the preparation of special urethane elastomer [10] and rigid forms [11]. Surprisingly the polymers were inert to boiling acid and alkali as well as certain organic solvents. However, amines were reported to convert the carbodiimides to guanidines as shown in Eq. (5).

$$\text{OCN}-\text{R}-\text{NCO} \xrightarrow{\underset{C_2H_5}{\overset{CH_3}{\underset{P}{\bigcirc}}}\text{O}} \{R-N=C=N\}_n + CO_2 \quad (4)$$

$$\{R-N=C=N\}_n + nR'NH_2 \longrightarrow \left[\begin{array}{c} R-NH-C=N \\ | \\ NHR' \end{array} \right]_n \quad (5)$$

Since the high molecular weight polycarbodiimides are insoluble and thus difficult to process, techniques to overcome this problem have been studied. One such approach involves adding given amounts of monofunctional isocyanates to give lower molecular weight polymers [12]. The rate of reaction was followed by carbon dioxide evolution with an American Meter Co. gas meter, and by infrared measurements of the isocyanate absorption at 2270 cm^{-1} and the carbodiimide peak at 2130 cm^{-1} [12]. After drying the polymer the DTA and TGA thermograms were obtained (DuPont 9N Thermal Analyzer) [12]. The number-average molecular weights were determined using a Hewlett-Packard vapor-phase osmometer [12]. The melt-processing viscosities were determined using a Brabender Plasti-Corder [12].

21-2. Polymerization of 2,4-Toluene Diisocyanate [13]*

(a) Preparation of catalyst: 1-Ethyl-3-methyl-3-phospholene 1-oxide.

$$C_2H_5PCl_2 + \underset{CH_2}{\overset{CH_2}{\diagdown}}\!\!\!\diagup^{CH_3} \longrightarrow C_2H_5\!\!-\!\!\underset{Cl}{\overset{Cl}{P}}\diagdown\!\!\diagup^{CH_3} \quad (6)$$

$$\downarrow H_2O$$

$$\underset{C_2H_5}{\overset{O}{P}}\diagdown\!\!\diagup^{CH_3}$$

A 3 liter, four-necked flask is fitted with a spiral condenser topped with a dry-ice condenser, thermometer, 1 liter dropping funnel with pressure equalizing side arm, and a magnetic stirrer. To this flask is added 1 gm of copper stearate, 780 gm (5.96 moles) of dichloroethylphosphine, and by the dropping funnel 447 gm (6.50 moles) of freshly distilled isoprene. The reaction mixture is stirred and refluxed under nitrogen for 42 hr, cooled and allowed to stand for 2

---

*Reprinted from T. W. Campbell, J. J. Monagle, and V. S. Foldi, *J. Am. Chem. Soc.* **84**, 3673 (1962). Copyright by the American Chemical Society. Reprinted by permission of the copyright owner.

days, and then refluxed without stirring for 5 days. Excess isoprene is distilled from the mixture and 850 ml of water is added dropwise with stirring to the reaction flask, which is cooled in an ice bath. The dark-brown, aqueous solution is transferred to a 5 liter flask and 1250 ml of 30% sodium hydroxide solution is added gradually to make the solution slightly alkaline (pH 8). The mixture is filtered and the aqueous solution is extracted continuously with chloroform for 12 days. The chloroform is distilled and the residue vacuum distilled through a 25 cm Vigreux column to give 435 gm (51%) of water-white liquid with a slight odor of phosphine. The product is further purified by oxidation at 50°C with excess 3% hydrogen peroxide for 6 hr. The aqueous mixture is extracted continuously with benzene and the oxide was recovered by distillation, b.p. 115°–119°C (1.2–1.3 mm), $n_D^{2.5}$ 1.5050.

Analysis
Calculated for $C_7H_{12}OP$: C, 58.4; H, 9.0; P, 21.5.
Found: C, 58.3; H, 8.8; P, 21.6

(b) *Polymerization of 2,4-toluene diisocyanate* [12][†].

$$CH_3\text{-}C_6H_3(NCO)_2 \xrightarrow[-2CO_2]{\underset{C_2H_5}{\overset{O}{\|}}P\text{-cyclic}} \left[-CH_3\text{-}C_6H_3\text{-}N=C=N\text{-}C_6H_3\text{-}CH_3-\right]_n$$

(7)

2,4-Toluene diisocyanate was distilled through a spinning band column. After a small forecut, distillation proceeded smoothly at 81°C (1.3 mm Hg). Polymerization of a 10% solution in boiling decahydronaphthalene was carried out with catalytic quantities of 1-ethyl-3-methyl-3-phospholene oxide. The reaction was complete in less than 1 hr, and the polymer was obtained in small, fluffy particles very reminiscent of puffed cereal in appearance and tex-

---

[†]Reprinted from T. W. Campbell and K. C. Smeltz, *J. Org. Chem.* **27**, 2069 (1962). Copyright by the American Chemical Society, 1962. Reprinted by permission of the copyright owner.

ture. These little particles were white and gave very tough, clear, nearly colorless film when pressed at 275°C. Strips of this film could be cold drawn; however, the film strips relaxed in boiling water and exhibited no crystallinity and extremely low X-ray orientation.

Analysis
Calculated for $(C_8H_6N_2)_x$: C, 73.8; H, 4.61; N, 21.4
Found: C, 73.4; H, 4.4; N, 21.0
                             73.5   4.2   21.0

3. POLYMERIZATION OF POLYISOCYANATES TO POLYISOCYANURATES

Hofmann [14] in 1858 was the first to report the trimerization of isocyanates to isocyanurates. Hofmann used triethylphosphine as a catalyst to trimerize phenylisocyanate to triphenylisocyanurate (Eq. 8). Diisocyanates would yield polymers with connecting isocyanurate rings [22].

$$3 \, C_6H_5NCO \xrightarrow{(C_2H_5)_3P} \text{triphenylisocyanurate} \quad (8)$$

Other trimerization catalysts that have been reported are amines [15, 16], salts of weak acids [17], potassium acetate, sodium carbonate [18], soluble compounds of K, Mg, Zn, Hg, Cu, Ni, Cr, Sn, Al, V, Ti [19], and epoxides in the presence of tertiary amines [20] such as triethylene diamine–propylene oxide [21]. Sandler [22] employed a variety of catalysts including calcium naphthenate, amines, and organometallics to yield polyisocyanates from diisocyanates and their prepolymers.

REFERENCES

1. A. J. Bur and L. J. Fetters, *Chem. Rev.* **76**, 727 (1976).
2. V. E. Shasona, W. Sweeney, and R. F. Tietz, *J. Am. Chem. Soc.* **82**, 866 (1960).
3. V. E. Shasona, *Macromol. Synth.* **1**, 63 (1963).
4. Y. Iwakura, K. Uno, and K. Ichikawa, *J. Polym. Sci., Part A-1* **2**, 3387 (1964); **6**, 793, 1087, and 2611 (1968).
5. C. King, *J. Am. Chem. Soc.* **86**, 437 (1963); W. L. Miller and W. B. Black, *Polym. Prepr., Am. Chem. Soc., Div. Polym. Chem.* **3**(2), 345 (1962).

6. R. G. Beaman, U.S. Patent 3,048,566 (1962).
7. T. W. Campbell, U.S. Patent 2,941,966 (1960); T. W. Campbell and J. J. Verbanc, U.S. Patent 2,853,473 (1958).
8. T. W. Campbell and K. C. Smeltz, *J. Org. Chem.* **27**, 2069 (1963); T. W. Campbell, J. J. Monagle, and V. S. Foldi, *J. Am. Chem. Soc.* **84**, 3673 (1962).
9. W. B. McCormack, *Org. Synth., Collect. Vol.* **5**, 787 (1973).
10. P. Fischer, W. Kallert, H. Holtschmidt, and E. Meisert, German Patent 1,145,353 (1963); A. Reischl, H. Holtschmidt, and P. Fischer, German Patent 1,143,018 (1963); General Tire & Rubber, British Patent 1,332,607 (1973).
11. H. None, R. Platz, and E. Wegner, Belgian Patent 657,835 (1964); P. Kan, C. Moses, and T. Narayan, U.S. Patent 3,772,217 (1973).
12. L. M. Alberino, W. J. Farrissey, Jr., and A. A. R. Sayigh, *J. Appl. Polym. Sci.* **21**, 1999 (1977).
13. T. W. Campbell, J. J. Monagle, and V. S. Foldi, *J. Am. Chem. Soc.* **84**, 3673 (1962).
14. A. W. Hofmann, *Jahresber. Fortschr. Agrikulturchem.* **1**, 349 (1858).
15. J. Burkus, U.S. Patent 2,993,870 (1961).
16. J. Burkus, U.S. Patent 2,979,485 (1961).
17. W. Frentzel, *Ber. Dtsch. Chem. Ges.* **21**, 411 (1888).
18. A. W. Hofmann, *Ber. Dtsch. Chem. Ges.* **18**, 764 (1885).
19. H. Havekoss, "Polymerization of Diisocyanates," O.P.B. Rep. 73894. France, 1942.
20. I. I. Jones and N. G. Savill, *J. Chem. Soc.* p. 4392 (1957); E. K. Moss, U.S. Patent 3,799,896 (1974).
21. B. D. Beitchman, *Rubber Age* **98**, No. 2, 65 (1966).
22. S. R. Sandler, *J. Appl. Polym. Sci.* **11**, 811 (1967).

22
POLYOXYALKYLATION OF HYDROXY COMPOUNDS

The purpose of this chapter is to describe the polyoxyalkylation of various alcohols and phenols. These compositions are used as polyurethane starting materials (polyols), as surfactants, and as textile aids.

Safety precautions. Ethylene oxide is a low boiling, flammable liquid, whose vapors easily form explosive mixtures in air [1]. It is to be labeled as a cancer hazard and a reproductive hazard [2]. It is prudent to handle propylene oxide and other three-membered ring organic oxides with safety precautions similar to those used for ethylene oxide.

The permissible 8 hr time weighted average exposure limit for ethylene oxide is 1 part per million parts of air [3]. Even at concentrations as "high" as 50 ppm, full-face respirators with EO-approved canister are required [4].

Work should be carried out in well-ventilated hood areas behind safety shields. Full protection against the explosive hazard associated with this compound is not achieved by this. It also should be kept in mind that the density of ethylene oxide is greater than that of air and, therefore, may not necessarily be carried up and out of the hood. Also, the problem of air pollution downwind from the hood's vent needs to be addressed. Special precautions must also be taken for the problems that may arise should there be a release of high levels of ethylene oxide as a result of an explosion.

Based on S. R. Sandler and W. Karo, "Polymer Syntheses," 2nd ed., Vol. III, pp. 154ff, by permission of Academic Press, San Diego, 1994.

It should be emphasized that ethylene oxide and propylene oxide are highly flammable. Contact with sources of ignition (sparks, electric heating elements, open flames, etc.) must be eliminated. The recommendations of the supplier's MSDS should be followed closely and considered as a minimum set of recommendations.

The exhaust gases from a preparation may be toxic and flammable. They may form explosive mixtures with air. Therefore, attention must be paid to the methods of venting the reaction vessel and attached peripherals both within the laboratory or plant and out-of-doors. Similar considerations also apply to the handling and disposal of the reaction mixtures, by-products, solvents, etc.

The physiological properties of polymers of ethylene oxide or propylene oxide or their derivatives are not well known. Presumably their solvent and/or surfactant properties may lead to unanticipated effects. For example, a compound of 1 mole of nonylphenol that has been reacted with 9 moles of ethylene oxide is a human spermicide used in certain contraceptives such as "Nonaxanol." Under other trade names, this compound finds application as a nonionic surfactant. Its analogue with 10 ethylene oxide units is the well-known "Triton X-100."

Despite its widespread use in consumer products, ethylene glycol, which is formed from ethylene oxide, is considered to be toxic. On the other hand, propylene glycol derivatives are thought to be of low or no toxicity.

Several companies have published descriptive bulletins for the safe handling of ethylene oxide, propylene oxide, and butylene oxide. Typical glassware and technical reviews are given in [5, 6].

1. CHEMISTRY

In general, most active hydrogen-containing compounds can be made to react with ethylene oxide (or propylene oxide) to give hydroxyethyl (or propyl) derivatives and on further reaction to give substituted polyoxyalkylene ethers as described in Eq. (1).

$$RZH + \underset{O}{\triangle}\!\!-\!R' \longrightarrow RZCH_2-\underset{|}{\overset{R'}{C}}H-OH$$

$$\underset{O}{\overset{\triangle-R'}{\longrightarrow}} RZ\left(CH_2-\underset{|}{\overset{R'}{C}}H-O\right)_n H \quad (1)$$

where Z = O [7], S [8], COO [9], NH [10]; R′ = H or CH_3; R = alkyl [11], aryl [11], H [12]. As a result of this reaction a mixture ($n = 1-100$) of polyglycol ethers may be obtained.

The oxyalkylation reaction can be catalyzed by either acids or bases. Some common catalysts are alkali and alkaline earth oxides, hydroxides and alkoxides, amines, tertiary amines, sodamide, zinc oxide, boron trifluoride, stannic chloride, and sulfuric acid. The reactions can be carried out at atmospheric pressure or at 10-50 psig in order to shorten reaction times. The usual temperatures employed are 100°-200°C for base-catalyzed reactions and 50°-70°C when Lewis acids are used.

The reaction is exothermic evolving about 20 kcal/mole of ethylene oxide reacted. Therefore adequate cooling capacity is essential to maintain reproducible (not runaway) reactions. At the end of the reaction, the catalyst is neutralized and the product is isolated.

The reactivity of alcohols and phenols with ethylene oxide varies in the order primary > alkyl phenol > secondary ≫ tertiary [13]. However, tertiary alcohols can be made to react with ethylene oxide [14]. Some linear secondary alcohols have been reacted with ethylene oxide to give nonionic surfactants (Union Carbide) [15]. Some typical alcohols used to give the nonionic surfactants (polyoxyethylenealkyl ethanols) are allyl, lauryl, cetyl, stearyl, tridecyl, myristyl, C_{12}-C_{15} primary linear, tallow, and trimethylnonyl [16].

Propylene oxide reacts with water using basic catalysts to give polyoxypropylene glycols [17]. In the case of primary and secondary aliphatic alcohols propylene oxide reacts to give monoalkyl ethers of polyoxypropylene glycol [18]. Propylene glycol reacts with propylene oxide to give polyoxypropylene glycols [19]. The use of triols give polyoxypropylene triols and pentaerythritol gives polyoxypropylene tetrols [20]. In a similar manner sucrose [19, 21] may be reacted with propylene oxide to give polyoxypropyl derivatives of sucrose. These liquid polyoxyalkylated alcohols are widely used as starting materials for the preparation of polyurethane foams and elastomers.

Particularly interesting block-copolymers are available from the reactions of ethylene oxide (monomer or polymers) with propylene oxide (monomer or polymers) or of 1,2-diaminoethane with ethylene oxide and/or propylene oxide. The series of surfactants in which a central polypropylene oxide is attached to polyethylene oxide polymeric chains are called Pluronic®; the series with a central polyethylene oxide surrounded by polypropylene oxides are called Pluronic® R. Products of diaminoethane reacted first with propylene oxide and terminated with ethylene oxide polymers are Tetronics. In the Tetronic® R series the arrangement of the oxides is reversed—the four terminal groups are polypropylene oxides while polymeric ethylene oxides are used to replace the four reactive hydrogens of 1,2-diaminoethane.

The ethylene oxide portions of the products confer hydrophilic characteristics on the molecules, the propylene oxide units are considered hydrophobic. By control of the molecular weights of the polymeric oxides, liquids, semisolids, and powders are formed. By the use of appropriate proportions of oxides, products with a wide range of application are available. Among these are uses that require defoaming/antifoaming properties. There are applications in cleaning; formation of foams, emulsions, and gels; solubilization; lubrication; wetting, thickening; and controlled dissolution [22].

In addition, phenols and naphthols are oxyalkylated to give products with varying units of either ethylene oxide or propylene oxide. Such compositions are commercially available as nonionic detergents.

Several reviews are worth consulting for additional information on the various aspects of the preparation and uses of polyoxyalkylene compounds [23].

The products resulting from the reaction of ethylene oxide with water or alcohols (or diols, etc.) are variously called poly(ethylene oxide) resins, poly(ethylene glycols), polyoxyethylene resins, PEOs, PEGs. The reaction products involving propylene oxide are the higher homologues known as poly(propylene oxide) resins, etc. The very high molecular weights PEGs have been called "POLYOX®."

With ordinary reaction catalysts such as conventional acids or bases, the products of the alkoxylation reactions are mixtures with a considerable range of molecular weights. By the appropriate adjustment of reactant ratios and reaction conditions, some control over the molecular weight of these products may be exercised. By this method, molecular weights from approximately 200 to 20,000 are obtained. By use of coordinate anionic initiator systems such as some involving ferric chloride, for example, high molecular weights from about 100,000 to 4,000,000 have been obtained.

With fluorinated Lewis acids, such as boron trifluoride, crown ethers rather than linear polymers may be formed [24].

On some catalyst surfaces, such as solid potassium hydroxide or alkaline earth compounds, PEGs or PPGs form. Since the molecular weights of the products from these reactions do not increase with conversion, it has been postulated that here we are dealing with a chain reaction rather than a polycondensation [25].

2. POLYOXYALKYLATION OF ALCOHOLS AND DIOLS

Mono- and dihydroxy compounds can be polyoxyalkylated with ethylene oxide, propylene oxide, ethylene carbonate, or propylene carbonate (Eq. 2).

$$ROH + \text{(epoxide)} \longrightarrow ROCH_2CH(R')-OH \longrightarrow RO\!\left[CH_2CH(R')-O\right]_n\!H \quad (2)$$

In most cases the reaction rate is increased by the use of an acidic or basic catalyst.

The advantage of alkylene carbonates is that the reaction can be run at atmospheric pressure.

In general, alcohols are reported to react faster than phenols and these in turn are reported to react faster than carboxylic acids [26].

The mono adducts are also obtained by the use of acid catalysts such as BF_3 etherate in the addition of ethylene oxide to primary, secondary, and tertiary alcohols [27]. Metal alkyls, particularly when used in binary systems, lead to high molecular weights and high conversions [28].

The initiation of the polymerization of alkylene oxides with iron has been known for some time. A patent extends this concept to the alkoxylation of many active hydrogen compounds. The catalyst for this process is a polycrystalline iron oxide (α-iron(III)oxide). For example, in an autoclave, under nitrogen, n-decanol and 2 wt % of the iron catalyst are treated with ethylene oxide at a pressure of less than 6 bars for 4 hr. The resulting liquid had a degree of polymerization of 5 [29].

22-1. Preparation of Ethyleneglycol Monoethyl Ether [27]

$$C_2H_5OH + \underset{(excess)}{CH_2\overset{O}{-}CH_2} \longrightarrow C_2H_5O-CH_2-CH_2OH \quad (3)$$

To an ice-cooled autoclave are added 10–30 moles of absolute alcohol and 2.0 mole of ethylene oxide. The autoclave is sealed and heated to 150°C for 12 hr or to 200°C for 3–4 hr. The autoclave is cooled and opened. The product is fractionally distilled to give 70% of ethylene glycol monoethyl ether, b.p. 134°C.

A more recent process [30] uses sodium methoxide as catalyst and the weight ratio of methanol to ethylene oxide is 9 : 1 to give 91.5% yield of ethyleneglycol monomethyl ether.

22-2. Polyoxypropylation and Polyoxyethylation of 2,6,8-Trimethyl-8-nonanol [31]

$$\begin{aligned}CH_3-\underset{}{\overset{CH_3}{\underset{|}{CH}}}-CH_2-\underset{OH}{\overset{CH_3}{\underset{|}{CH}}}-CH_2-\overset{CH_3}{\underset{|}{CH}}-CH_2-\overset{CH_3}{\underset{|}{CH}}-CH_3 + \overset{CH_3}{\underset{O}{\triangle}} \xrightarrow{BF_3} \\ \xrightarrow[NaOH]{\triangle\!O} CH_3-\overset{CH_3}{\underset{|}{CH}}-CH_2-\underset{O(CH_2-\underset{CH_3}{\underset{|}{CHO}}-)_x(-CH_2CH_2O)_yH}{\overset{CH_3}{\underset{|}{CH}}}-CH_2-CH-CH_2-\overset{CH_3}{\underset{|}{CH}}-CH_3 \quad (4)\end{aligned}$$

To a 1 liter, four-necked flask equipped with an agitator, thermometer, addition funnel, and condenser are added 186 gm (1.0 mole) of 2,6,8-trimethyl-4-nonanol, and 0.55 gm (0.03 mole) of boron trifluoride. The reaction mixture is purged with nitrogen and then 116 gm (2.0 moles) propylene oxide; is added over a 1 hr period at 75°–80°C with constant agitation. The acid catalyst is neutralized with methanolic caustic to pH 8.5. Then 0.3 gm sodium hydroxide is added and the temperature raised to 150°–160°C; ethylene oxide is added (369 gm = 8.4 mole) at 0–15 lb pressure to a cloud point (1% in water) at 36°C. The final product has surfactant properties.

A similar two-stage ethoxylation of secondary alcohols of 10–17 carbon atoms has been described by Carter [32].

22-3. Preparation of Polyoxypropylated Stearyl Alcohol [33]

$$C_{18}H_{37}OH + CH_2\underset{O}{\overset{CH_3}{\underset{\diagdown\diagup}{-}CH}} \longrightarrow C_{18}H_{37}O\left(CH_2-\overset{CH_3}{\underset{|}{CH}}-O\right)_5 H \quad (5)$$

To a 1 liter stainless steel autoclave are added 230 gm (0.8 mole) of stearyl alcohol and 0.9 gm of sodium methoxide powder. The reactor is purged with nitrogen. Then 408 gm (7.0 mole) of propylene oxide is added over a 16 hr period while the temperature is kept at 130°-135°C. After the reaction is complete the mixture is cooled and the volatiles are eliminated by heating on a water bath for 15-20 min. Then dilute sulfuric acid is added to neutralize the catalyst and the product is washed with 1 liter of water at 60°C, followed by two washings with 500 ml of water at 60°C. The product is dried under reduced pressure to give 548 gm (99%) of a light-yellow product having a hydroxyl number of 94.

22-4. Preparation of Polyoxypropylene Glycol [19]

$$CH_3-\underset{OH}{\overset{}{\underset{|}{CH}}}-\underset{OH}{\overset{}{\underset{|}{CH_2}}} \xrightarrow[\text{NaOH}]{\overset{CH_3}{\underset{\diagdown\diagup}{\overset{|}{CH_2}}}} HO\left(CH_2-\overset{CH_3}{\underset{|}{CH}}-O\right)_n H \quad (6)$$

To a 1 liter, three-necked, round-bottom flask equipped with a mechanical stirrer, reflux condenser, thermometer, and additional funnel is added 57 gm (0.75 mole) of propylene glycol and 7.5 gm (0.19 mole) of sodium hydroxide. The flask is purged with nitrogen to remove air and heated to 120°C with stirring to dissolve the sodium hydroxide. Then propylene oxide is added (40-45 moles). The reaction mixture is cooled under nitrogen, neutralized with dilute sulfuric acid, filtered, and dried under reduced pressure to give a water-insoluble product with MW 1620 as determined by hydroxyl number or acetylation analytical procedures.

Among new initiators for the polymerization of propylene oxide, ethylene oxide, 1-butene oxide, and isobutylene oxide are zinc hexacyanocobaltate [34] and aluminum porphyrin [35].

It is of interest that when aluminum porphyrin is used, oligomers or polymers of controlled molecular weight distribution can be produced by using bisphenol A or related bisphenol-type compounds as chain transfer agents.

Living cationic polymerizations of propylene oxide and of epichlorohydrin have been reported. The procedure is carried out in an alcohol with a strong acid catalyst such as fluoroboric acid. Interestingly enough, the procedure is not applicable to the polymerization of ethylene oxide [36].

3. POLYOXYALKYLATION OF POLYHYDROXY COMPOUNDS

The polyoxyalkylation of polyhydroxy compounds is usually carried out using basic catalysts in high-pressure reactors. The starting polyols are in most cases solids and are reacted with other ethylene oxide or propylene oxide with or without a solvent (e.g., water, xylene, DMSO), under about 100 psig pressure at temperatures ranging from 100° to 150°C. In most cases the reaction is run under anhydrous conditions.

22-5. Polyoxypropylation of Sucrose [37]

where R = —$CH_2CH(CH_3)$—.

To a steam-heated autoclave are added 950 gm (28.00 moles) of sucrose, 3 gm of 1,2,4-trimethylpiperazine catalyst, and 60 ml of distilled water. The autoclave is purged three times with nitrogen gas and heated to 100°C. Then the addition of 1550 gm (27.0 mole) of propylene oxide is started. After 1 hr the temperature is raised to 115°C. At this temperature the addition is carried out at approximately 90 psig. The addition of the remaining propylene oxide requires about 10.25 hr. The reactants are then heated to 120°C for another 45 min. The reaction mixture is cooled to 60°C and the liquid product is blown into a clean vessel. The volatiles are removed under reduced pressure varying from 80 mm Hg to a final 2 mm Hg over a 5 hr period. The product has MW 839.

This product can be used again in a second stage of polyoxypropylation wherein it acts as a solvent for fresh sucrose. The reaction conditions are basically as described before. Using this procedure 670 gm (0.8 mole) of sucrose-propylene oxide product is used for 670 gm (1.95 mole) of fresh sucrose.

22-6. Polyoxyethylation of Sorbitol [38]

$$\begin{array}{c} CH_2OH \\ | \\ (CH-OH)_4 \\ | \\ CH_2OH \end{array} + CH_2\!\!-\!\!\!\!\underset{O}{\diagdown\!\!\diagup}\!\!\!\!-\!CH_2 \longrightarrow \begin{array}{c} CH_2O\!\!-\!\!(CH_2CH_2O)_n\!H \\ | \\ (CH-O\!\!-\!\!(CH_2CH_2O)_n\!H)_4 \\ | \\ CH_2O\!\!-\!\!(CH_2CH_2O)_n\!H \end{array} \quad (8)$$

To 200 gm sorbital in an autoclave is added 350 gm ethylene oxide. The mixture is stirred and heated for 12 hr at 140°C. The reaction mixture is cooled and the excess ethylene oxide removed to give a viscous water-white liquid.

The authors of the foregoing procedure report that boric or sulfuric acid facilitate the polyoxyethylation reaction. However, other researchers report that basic catalysts such as sodium methoxide are preferred [39–41].

The use of ethylene carbonate has the advantage of allowing the reaction to be carried out at atmospheric pressure in typical laboratory glassware.

22-7. Preparation of Polyoxyethylated Pentaerythritol [42]

$$\begin{array}{c} CH_2OH \\ | \\ HO-CH_2-C-CH_2OH \\ | \\ CH_2OH \end{array} + \begin{array}{c} O \\ \| \\ C \\ / \ \backslash \\ O \quad O \\ | \quad | \\ CH_2-CH_2 \end{array} \xrightarrow{K_2CO_3}$$

$$C[-CH_2O(-CH_2-CH_2O)_n H]_4 \quad (9)$$

To a glass resin-flask equipped with a stirrer and condenser are added 408 gm of pentaerythritol, 528 gm ethylene carbonate, and 10 gm of potassium carbonate. The temperature is slowly raised to 155°C at which point the mixture begins to evolve carbon dioxide and become homogeneous. Heating and stirring is continued for 6 hr and the temperature is slowly raised to 200°C. The product is a dark, viscous liquid that is soluble in water or in a xylene–methanol mixture.

The polymerization of ethylene oxide and propylene oxide with various active hydrogen compounds is the primary thrust of the present chapter. However, other epoxy compounds also undergo this type of reaction (cf [43]).

4. POLYOXYALKYLATION OF PHENOLS

Phenol and its O-alkyl derivatives (from propylene dimers, trimers, or tetramers) are oxyalkylated usually at atmospheric pressure or under moderate pressure conditions using basic catalysts at approximately 50°–200°C in the absence of solvents. Glass reaction flasks are the typical labware, thus the reaction can be carried out conveniently in the laboratory. However, in the absence of catalysts the reactions require high temperatures and pressures [44–47].

The products have been used since the early 1940s as nonionic surfactants and appear under a variety of trade names.

The reactivity of the various phenols depends on the substituent present. Reactivity is increased if the substituent decreases the acidity of the phenol (electron-donating groups) and is decreased if the acidity is increased (electron-withdrawing groups) [45].

The reaction of ethylene oxide with phenol in the presence of sulfuric acid gives o-vinylphenol [47].

22-8. Preparation of Polyoxyethylated Nonylphenol [48]

$$C_9H_{19}{-}\!\!\bigcirc\!\!{-}OH + CH_2\!\!-\!\!CH_2 \xrightarrow{NaOH}$$
$$\phantom{C_9H_{19}{-}\!\!\bigcirc\!\!{-}OH + CH_2}\underset{O}{\diagdown\!\diagup}$$

$$C_9H_{19}{-}\!\!\bigcirc\!\!{-}O{-}(CH_2CH_2O)_{10}H \quad (10)$$

To a reactor are charged 890 lb of monononylphenol and 404 gm sodium hydroxide. The reaction mixture is heated to 140°C and then 1630 lb ethylene oxide is added over a 6 hr period at 140°–180°C while the pressure remains at 40 psig. At the end of the reaction 350 gm of 85% H_3PO_4 is added to neutralize the sodium hydroxide.

22-9. Polyoxypropylation of p-tert-Butylphenol [48]

$$(CH_3)_3C{-}\!\!\bigcirc\!\!{-}OH + CH_2\!\!-\!\!\overset{CH_3}{\underset{}{CH}} \xrightarrow{NaOH}$$

$$(CH_3)_3C{-}\!\!\bigcirc\!\!{-}O{-}\!\!\left[CH_2{-}\overset{CH_3}{\underset{}{CH}}{-}O\right]_{2-5}\!\!H \quad (11)$$

To a 2-liter, three-necked, glass, round-bottom flask equipped with a mechanical stirrer, thermometer, addition funnel, and dry ice–acetone condenser is added 600 gm (4.0 mole) of p-tert-butylphenol. The flask is heated in order to melt the latter phenol and then 8.0 gm (0.2 mole) of sodium hydroxide is added. The temperature is raised to 150°C and then 929 gm (16 moles) of propylene oxide is added in 50 ml increments over a 7.5 hr period while slowly raising the temperature to 225°C during this period. Heating is continued. After cooling to room temperature the product is neutralized with a solution of 20.8 gm conc HCl in 150 ml water. The organic layer is separated and washed with 500 ml saturated salt solution. This product is filtered, dried and distilled under reduced pressure to give four products:

$$(CH_3)_3C-\underset{}{\underset{}{\bigcirc}}-O-\left(CH_2CH-O\right)_m H$$
$$\phantom{(CH_3)_3C-\underset{}{\underset{}{\bigcirc}}-O-(}CH_3$$

m	b.p. (°C)	p (mm Hg)	Specific gravity	$[n]_D^{20}$
2	121–137	0.15–0.45	0.9924	1.4968
3	153–168	0.2–0.6	0.9929	1.4888
4	181–196	0.2–0.4	0.9970	1.4819
5	198–222	0.25–0.65	0.9967	1.4752

REFERENCES

1. Code of Federal Regulations. Labor, Vol. 29, Ch. XVII (7-1-90 Edition), Part 1910, p. 303.
2. Ref. [1], p. 300.
3. Ref. [1], p. 295.
4. Ref. [1], p. 298.
5. S. R. Sandler and W. Karo, "Polymer Syntheses," 2nd ed., Vol. 1, Chapter 6, Academic Press, San Diego, 1992.
6. "Alkylene Oxides," Technical Bulletin 125-551-65, Dow Chemical Company, Midland, Michigan, 1965; M. M. Padwe, "The Preparation of Nonionic and Other Surface-Active Agents Based on Ethylene Oxide," Literature Survey, Vol. 1, Publication R-Ab 13, Jefferson Chemical Company, Houston, Texas, 1951; "Ethylene Oxide," Technical Brochure J2-024-5-7-62, Jefferson Chemical Company, Houston, Texas; "Alkylene Oxides," Technical Bulletin F-40558, Union Carbide Chemical Company; "Ethylene Oxide," Technical Bulletin 1500-5-61, GAF Corporation; "Technical Data Sheet on Ethylene Oxide," 0-622, Wyandotte Chemical Company, Wyandotte, Michigan.
7. D. R. Jackson and L. G. Lundsted, U.S. Patent 2,677,700 (1954); R. E. Hefner and M. E. Pruitt, British Patent 813,495 (1959); F. H. Roberts and H. R. Fife, U.S. Patent 2,425,755 (1947); C. Schaller and N. Wittwer, U.S. Patent 1,970,578 (1934).
8. H. Schuette, C. Schöeller, and M. Wittwer, U.S. Patent 2,129,709 (1938); W. A. Schulze et al., Ind. Eng. Chem. 42, 920 (1950); E. W. Gluesenkemp, U.S. Patent 2,498,617 (1950).
9. J. T. Patton, Jr., U.S. Patent 3,101,374 (1963); H. Schuette and M. Wittwer, U.S. Patent 2,174,760 (1939).
10. M. DeGroote, U.S. Patents 2,626,902 and 2,626,912–919 (1953).
11. W. D. Harris and J. W. Zukel, U.S. Patent 2,820,808 (1958).
12. P. H. Schlosser and K. R. Gray, U.S. Patent 2,362,217 (1944).
13. G. J. Stockburger and J. D. Brandner, J. Am. Oil Chemist's Soc. 40, 590 (1963).

14. W. Umbach and W. Stein, U.S. Patent 3,651,152 (1972).
15. J. H. McFarland and P. R. Kinkel, *J. Am. Oil Chemist's Soc.* **41**, 742 (1964).
16. C. E. Stevens, in "Kirk–Othmer Encyclopedia of Chemical Technology" (A. Standen et al., eds.), 2nd ed., Vol. 19, pp. 538–539, Wiley (Interscience), New York, 1969.
17. P. H. Schlosser and K. R. Gray, U.S. Patent 2,362,217 (1944).
18. R. E. Hefner and M. E. Pruitt, British Patent 813,495 (1959); F. H. Roberts and H. R. Fife, U.S. Patent 2,425,755 (1947); D. R. Jackson and L. G. Lundsted, U.S. Patent 2,677,700 (1954).
19. L. G. Lundsted, U.S. Patent 2,674,619 (1954).
20. L. T. Monson and W. J. Dickson, U.S. Patent 2,766,292 (1956).
21. M. DeGroote, U.S. Patent 2,552,528 (1951).
22. "Pluronic and Tetronic Surfactants," BASF Corporation, Specialty Products, Parsippany, New Jersey, 1989.
23. D. B. Braun and D. J. DeLong, in Kirk–Othmer Encyclopedia of Chemical Technology," 3rd ed., Vol. 18, p. 616, Wiley (Interscience), New York, 1980; R. A. Newton, in "Kirk–Othmer Encyclopedia of Chemical Technology," 3rd ed., Vol. 18, p. 633, Wiley (Interscience), New York, 1980; N. Schönfeldt, "Surface Active Ethylene Oxide Adducts," Pergamon, Oxford, 1969; J. Furukawa and T. Saegusa, "Polymerization of Aldehydes and Oxides," Wiley (Interscience), New York, 1963; F. E. Bailey and J. V. Koleske, "Poly(ethylene oxide)," Academic Press, New York, 1976; A. S. Kasten and N. G. Gaylord (eds.), "Polyethers," Wiley (Interscience), New York, 1963; R. J. Ceresa (ed.), "Block and Graft Copolymerization," Wiley (Interscience), London/New York, 1973; W. J. Burlant and A. S. Hoffman, "Block and Graft Copolymers," Reinhold, New York, 1961; E. J. Vandenberg, "Catalysis: A Key to Advances in Applied Polymer Science," ACS Symposium Series, Vol. 496, 1992; M. Shibata, M. Saito, and S. Akimoto (eds.), "Poly(alkylene oxides): Manufacture, Characteristics, and Uses," Kaibundo, Tokyo, 1990; N. Clinton and P. Matlock, in "Concise Encyclopedia of Polymer Science and Engineering" (J. I. Kroschwitz, Ed.), p. 337, Wiley, New York, 1990; S. D. Gagnon, in "Concise Encyclopedia of Polymer Science and Engineering," p. 343, Wiley, New York, 1990; J. Furukawa and T. Saegusa, in "Encyclopedia of Polymer Science and Technology," Vol. 6, p. 175, 1967; F. E. Bailey, Jr., and J. V. Koleske, "Alkylene Oxides and Their Polymers," Dekker, New York, 1991; L. C. Pizzini and J. T. Patton, Jr., in "Encyclopedia of Polymer Science and Technology," Vol. 6, pp. 145–168, 1967; F. W. Stone and J. J. Stratta, in "Encyclopedia of Polymer Science and Technology," Vol. 6, p. 103, 1967; C. E. Stevens, in "Encyclopedia of Chemical Technology," Vol. 19, p. 531, 1969; A. M. Schwartz and J. W. Perry, "Surface Active Agents, Their Chemistry and Technology," Vol. I, pp. 202–217, Wiley (Interscience), New York, 1949.
24. "Kirk–Othmer Encyclopedia of Chemical Technology," 3rd ed., Vol. 9, p. 436, Wiley, New York, 1980.
25. L. E. St. Pierre and C. C. Price, *J. Am. Chem. Soc.* **78**, 3432 (1956); F. N. Hill, F. E. Bailey, Jr., and J. T. Fitzpatrick, *Ind. Eng. Chem.* **50**, 5 (1958).

26. C. Schaller and M. Wittwer, U.S. Patent 1,970,578 (1934); H. Schuette and M. Wittwer, U.S. Patent 2,174,760 (1940); R. D. Fine, *J. Am. Oil Chem. Soc.* **35**, 542 (1958); A. N. Wrigley, F. D. Smith, and A. J. Stirton, *ibid.* **34**, 39 (1957).
27. C. A. Carter, U.S. Patent 2,870,220 (1959); W. Umbach and W. Stein, U.S. Patent 3,651,152 (1972).
28. J. Furukawa, T. Saigusa, T. Tsuruta, and G. Kakogawa, *Makromol. Chem.* **36**, 25 (1959).
29. W. Hoelderich, J. Houben, G. Wolf, and M. G. Kinnaird, U.S. Patent 5,126,493 (1992).
30. G. Cocazza, B. Calcagno, and G. Torreggiani, U.S. Patent 3,935,279 (1976).
31. R. E. Leary, L. T. Nehmsmann, III, and L. M. Schenck, U.S. Patent 3,350,462 (1967).
32. C. H. Carter, U.S. Patent 2,870,220 (1959).
33. F. Lachampt and G. Vanderberghe, U.S. Patent 3,489,690 (1970).
34. H. Takeyasu, Y. Matsumoto, and O. Shigayuki, Japanese Patent 64,268,329[92,268,329], Feb. 25, 1991; S. L. Aggarwal, I. G. Hargis, and R. A. Livigni, U.S. Patent 4,279,798 (1980).
35. J. E. McGrath, *Chemtech*, May 1991, 310, and Refs. 28 and 29 therein.
36. O. W. Webster, *Science* **251**, 887 (1991).
37. J. T. Patton, Jr. and W. F. Schulz, U.S. Patent 3,346,557 (1967).
38. O. Schmidt and E. Moyer, U.S. Patent 1,922,459 (1933).
39. M. DeGroote, U.S. Patent 2,552,528 (1951).
40. J. T. Patton, Jr. and W. F. Schulz, U.S. Patent 3,346,557 (1967).
41. A. W. Anderson, U.S. Patent 2,927,918 (1960).
42. L. T. Monson and W. J. Dickson, U.S. Patent 2,766,292 (1956).
43. S. R. Sandler and J. M. Boben, U.S. Patent 4,393,248 (1983).
44. R. A. Smith, *J. Am. Chem. Soc.* **62**, 994 (1940); H. M. Stanley, J. E. Youell, and G. Minkoff, U.S. Patent 2,141,443 (1938); *Chem. Abstr.* **33**, 2536 (1939).
45. D. R. Boyd and E. R. Marle, *J. Chem. Soc.* **105**, 2117 (1914); J. M. Cloney and R. L. Mayhew, *Soap Chem. Spec.* **33**, No. 8, 52, 109, and 111 (1957).
46. R. L. Mayhew and R. C. Hyatt, *J. Oil Colour Chem. Soc.* **29**, 357 (1952); S. A. Miller, B. Bann, and R. D. Thrower, *J. Chem. Soc.* p. 3623 (1950).
47. R. A. Smith and J. B. Niederl, *J. Am. Chem. Soc.* **53**, 806 (1931).
48. W. D. Harris and J. W. Zukel, U.S. Patent 2,820,808 (1958).

23
POLYMERIZATION OF VINYL ESTERS

In this section, the emphasis is on the methods for polymerizing vinyl acetate. Other vinyl esters will only be mentioned briefly.

Besides applications of poly(vinyl acetate) (PVAc) homopolymers, there are major uses of copolymers of vinyl esters with such comonomers as vinyl chloride, ethylene, various acrylic and methacrylic esters, etc.

Poly(vinyl acetate) is used in floor tile, chewing gum bases, paper coatings, latex paints, adhesives, and in textile treatments. In the textile field the polymer goes into bodying and stiffening agents, binders for pigments, fabric sizes, bonding agents for nonwoven textiles, and as a material which improves the abrasion resistance of the substrate. Its durability, transparency, and stability to weathering and sunlight are attributes that contribute to its industrial acceptance.

A major fraction of the industrial homopolymer of vinyl acetate is converted to poly(vinyl alcohol). Since the monomer vinyl alcohol does not exist, poly(vinyl alcohol) must be produced by hydrolysis or alcoholysis of a polymeric vinyl ester. [See Chapter 25 for details on the preparation of poly(vinyl alcohol).]

While the vinyl esters of a large variety of acids have been studied, few have found significant industrial applications. For example, during the 1950s, vinyl stearate was produced in the United States. The intended application for this compound was as a polymerizable internal plasticizer for poly(vinyl acetate). However, the

From S. R. Sandler and W. Karo, "Polymer Syntheses," 2nd ed., Vol. III, pp. 202ff, by permission of Academic Press, San Diego, 1994.

cost of production included not only that of the vinylation of stearic acid, but also that of separating inhibiting oleic acid from the starting material. Consequently, this monomer could not compete either on the basis of cost or of plasticizing efficiency with ethylene or with the 2-ethylhexyl esters of maleic or acrylic acids. Vinyl propionate, whose polymerization characteristics are somewhat unique among the vinyl esters of fatty acids, and vinyl esters of mixed, branched acids in the C_9- and C_{10}-range ("Versatic Acids") have commercial applications [1, 2].

Quite recently, Union Carbide Corp. introduced "Vinate" monomers. These are vinyl pivalate, vinyl neodecanoate, vinyl neononanoate, vinyl propionate, and vinyl 2-ethylhexanoate. These monomers are said to improve the hydrolytic stability and water resistance of emulsion copolymers [3].

The literature on vinyl esters and their polymers is quite extensive albeit somewhat sketchy from the preparative standpoint. Among the leading reviews are Lindemann [1 (which cites 33 general references in its bibliography), 2], Bartl [4], Tromsdorff and Schildknecht [5], El-Aasser and Vanderhoff [6], Daniels [7, 8], and Rohde [9]. Two reviews about poly(vinyl alcohol) are by Cincera [10] and Marten [11].

Safety precaution. Traditionally, vinyl acetate monomer has been considered a flammable liquid which is also an irritant. At the present time, in the U.S., the toxicity status of all chemical intermediates is in a constant state of flux, and any statement made here may very well be out of date by the time this work reaches the reader. We therefore recommend that prior to handling the monomer, the current manufacturer's Material Safety Data Sheet (MSDS) be consulted.

In March 1993, an explosion, resulting in one fatality, occurred in a German plant which produced poly(vinyl acetate) and poly(vinyl alcohol). The exact cause of the explosion was not reported. However, the news article in *Chemical and Engineering News* stated that, at least in Germany, vinyl acetate is "under justified suspicion" for being a potential carcinogen [12]. In the light of this statement, prudence dictates the use of due precautions in the handling of vinyl acetate and other vinyl esters. Naturally precautions must also be taken to protect personnel and equipment against fires and explosions.

The polymers derived from vinyl acetate are considered relatively innocuous. Poly(vinyl alcohol) is thought to be harmless. The toy "Slime" is obtained by mixing a solution of poly(vinyl alcohol) with a borax solution [13]; some grades of poly(vinyl acetate) have been used in chewing gum bases.

Commercially produced vinyl acetate monomer is generally of high purity. Table I illustrates the effects on the polymerization of a variety of small amounts of impurities that may be found in samples of vinyl acetate.

1. MONOMER PURIFICATION

The purification procedures of vinyl acetate for precise kinetic studies are not entirely satisfactory. Many of the impurities mentioned in Table I, for example, may form azeotropic compositions with the monomer. Dissolved oxygen seems to be difficult to remove even on repeated degassing. Acetaldehyde, which is a significant chain-transfer agent, is particularly troublesome since it forms readily by hydrolysis of vinyl acetate. The separation of inhibitors by distillation is said to be difficult [16].

The techniques of purification consist of three procedures: (1) washing with appropriate reagents, (2) fractional distillation in an atmosphere from which oxygen has been rigorously excluded, and (3) partial bulk polymerization followed by distillation, or various combinations of these three methods. These procedures may then be followed by repeated careful degassing.

The effect of oxygen on the polymerization characteristics of the monomer is quite dramatic. In bulk polymerization of reasonably purified vinyl acetate, the process is autoaccelerated from the start and goes nearly to completion with a residual monomer content of 2–4%. When the monomer distillation is carried out in contact with air, inhibiting impurities form quite rapidly. These lead to dead-end polymerizations with 30–40% unreacted monomer left in the product [17].

A number of variations of the distillation procedure have been mentioned. For example, the monomer has been fractionally distilled through a 25-theoretical plate column and polymerized to approximately 10% conversion. Then the residue has been distilled directly into a dilatometer [18], the monomer has been distilled and percolated through a silica gel column prior to use [19]. In another

TABLE I
EFFECTS OF TYPICAL IMPURITIES ON THE POLYMERIZATION OF VINYL ACETATE[a]

Inhibitors
  Oxygen
  Hydroquinone, other phenolic compounds
  Phenothiazine
  Diphenylamine
Retarders
  Crotaldehyde (also acts as chain-transfer agent)
  Divinylacetylene
  Vinylacetylene
  Crotonic acid (also copolymerizes)
  Butadienylacetylene
  Isopropenyl acetate (also copolymerizes)
  Vinyl crotonate (also copolymerizes)
  Copper acetate
Chain-Transfer Agents
  Oxygenated products from the action of oxygen on vinyl acetate
  Acetic acid
  Acetone
  Crotonaldehyde (also acts as retarder)
  Methanol
  Methyl acetate
  Aromatic solvents such as benzene and toluene
Copolymerizing Impurities
  Methyl vinyl ketone
  Crotonic acid (also acts as a retarder)
  Isopropenyl acetate (also acts as a retarder)
  Vinyl crotonate (also acts as a retarder)
Impurities with no Significant Effect
  Water (up to 5%)
  Ethylidene diacetate (up to 5%)[b]

[a] From Ref. [14].
[b] In a polymerization study of vinyl benzoate, a definite inhibiting effect was attributed to ethylidene diacetate (1,1-diacetoxyethane) [15].

procedure, the monomer was distilled at 300 mm Hg in an oxygen-free nitrogen atmosphere through a 2 cm × 60 cm glass-helix-filled column, retaining only middle fractions boiling within 0.1°C of each other [20]. Inhibitor-free monomer was distilled three times at 300–400 mm Hg under nitrogen through a 25-theoretical plate column, a process during which partial copolymerization contributed to the elimination of impurities [21]. A reduced pressure distillation under argon using a 15-theoretical plate glass column has also been reported [22].

Purification procedures which depend primarily on the removal of interfering materials by partial polymerization of the monomer are usually not described in detail [16]. It appears that the monomer is generally warmed with a typical free-radical initiator until a slight increase in viscosity is observed. Then the unpolymerized monomer is distilled off in an inert atmosphere through a fractionating column. The monomer may be predried with a zeolite such as Linde 4A prior to partial polymerization and fractional distillation [23].

Gunesch and Schneider [24] prepolymerized a distilled sample of vinyl acetate in the presence of as much as 5–8% of lauroyl peroxide. When the extent of the polymerization had been judged to be sufficient, the process was "short-stopped" by the addition of 2,4-dinitrophenylhydrazine. This step, it seems to us, may also assist in the removal of traces of acetaldehyde. The monomer was then fractionally distilled in the usual manner through a column of 10 theoretical plates.

Those methods of purifying vinyl acetate which involve washing of the monomer usually seem to be concerned with the removal of hydroquinone, acetic acid, and acetaldehyde. The use of aqueous solutions for these procedures actually is somewhat questionable in view of the ease of hydrolysis of vinyl acetate (comparable in rate to the hydrolysis of ethyl acetate). In aqueous alkali, the hydrolysis rate of this monomer is said to be 370 times as fast as in water [2]. Nevertheless, such procedures have been suggested and are given here for information only as a typical example of this method.

The percolation of the monomer through a short column packed with an aluminum oxide such as Alcoa Alumina CG20 should be evaluated as a means of removing phenolic inhibitors and their oxidation products. This approach has been successful in styrene and acrylic ester deinhibition. It also has the virtue of reducing the water content of the monomer and the possibility of uncontrolled hydrolysis.

The older literature generally recommends the use of purified nitrogen to create an inert atmosphere in a reaction system. Since the density of nitrogen is close to that of air, nitrogen has to be bubbled into a reactor constantly to prevent oxygen from entering. Since argon is sufficiently high in density that it can form a much more satisfactory inert blanket in the system, we suggest that argon be substituted generally for forming a stable inert atmosphere in a reactor.

2. INITIATION, INHIBITION, AND RETARDATION OF POLYMERIZATION

The role of aromatic moieties in the polymerization of vinyl acetate is interesting since benzene, for example, may be considered a natural solvent for the solution of polymerization of vinyl esters. Yet, in the presence of oxygen, cumene (isopropylbenze) inhibits the autoxidation process which may use up oxygen. Thus, the onset of polymerization may be delayed. On the other hand, when oxygen has been stringently excluded from the system, cumene acts as a simple retarder.

All aromatic solvents, including those without alkyl side chains such as chlorobenzene, ethyl benzoate, and benzene, retard the polymerization of vinyl acetate [25].

It has been postulated that the very reactive free radical formed from vinyl acetate reacts with an aromatic nucleus to form a less reactive free radical capable of adding more monomer (Eq. 1).

$$R+CH_2-CH)_n-CH_2-CH \cdot + \bigcirc \longrightarrow$$
$$\quad\quad\quad\quad\quad\quad | \quad\quad\quad\quad |$$
$$\quad\quad\quad\quad\quad\quad O \quad\quad\quad\quad O$$
$$\quad\quad\quad\quad\quad\quad | \quad\quad\quad\quad |$$
$$\quad\quad\quad\quad\quad\quad C=O \quad\quad\; C=O$$
$$\quad\quad\quad\quad\quad\quad | \quad\quad\quad\quad |$$
$$\quad\quad\quad\quad\quad\quad CH_3 \quad\quad\; CH_3$$

$$R+CH_2-CH)_n-CH_2-CH-CH\begin{smallmatrix}CH=CH\\ \diagup\quad\quad\diagdown\\ \diagdown\quad\quad\diagup\\ CH=CH\end{smallmatrix}CH \cdot \quad (1)$$

In a study using ^{14}C-labeled benzene, it was found that at a degree of polymerization of 700, the average polymer molecule contained about 20 benzene units [25]. Thus there is evidence of retardation attributable to a degradative chain-transfer process [25, 26]. In addition, there is evidence that actual copolymerization of vinyl acetate and benzene takes place. As a matter of fact, there may be controversy whether the observations discussed here are a result of chain transfer or of copolymerization. The point in question may be rather subtle [27]. The facts, regardless of theoretical interpretation, are that aromatic solvents retard the polymerization of vinyl acetate, and the polymer contains substantial quantities of covalently bound solvent.

Related to these observations are studies on the polymerization of vinyl benzoate [28]. The monomer is an example of a self-inhibiting reactant. During polymerization, aromatic radical adducts form which are sufficiently stable to result in a retardation of the rate of polymerization as the copolymer with the benzene ring forms. During the process, it is also possible for the aromatic radical adduct to react with monomer to re-form the aromatic nucleus as well as form vinyl radicals.

With increasing temperatures, there is evidence that the mechanism of the polymerization gradually changes from a rate of polymerization proportional (as is conventional) to the square root of the initiator concentration at 60°C (using AIBN as initiator) to a rate directly proportional to the first power of the AIBN concentration at 95°C. Changes in the termination steps with temperature may explain these transitions [28].

Once the above observations have been made, questions naturally arise concerning the influence of the aromatic nuclei of such initiators as benzoyl peroxide on the polymerization process. Indeed chain transfer between benzoyl peroxide and poly(vinyl acetate) free radicals has been observed [29]. The copolymerization of benzoyloxy radicals with vinyl formate, vinyl propionate, vinyl butyrate, vinyl benzoate, and vinyl phenylacetate has been studied in considerable detail [30, 31].

3. POLYMER STRUCTURE

The nature of the end groups of high molecular weight polymers is a general problem in polymer chemistry. The analytical determination of a functional group with a molecular weight on the order of 10^2 in a molecule with a molecular weight in the range of 10^5 is complicated by the fact that ordinary polymers consist of chains with a wide distribution of molecular weights. The two ends of each chain probably have different compositions and the nature of the end groups varies with the nature of initiators, solvents, chain-transfer agents, surfactants, incidental impurities, etc. The chemical reactions of functional groups attached to polymer chains are not always well understood, e.g., conformational factors and shielding effects due to a coiling of a polymer chain around a potentially reactive site are two obvious matters that may interfere with conventional chemical procedures. For these and many other related reasons, end-group analyses for all polymeric systems must be viewed with considerable skepticism.

In the field of poly(vinyl acetate) research, it was noted early that hydrolysis of the polymer to poly(vinyl alcohol) followed by reacetylation led to the production of poly(vinyl acetates) of lower average molecular weight distribution than that of the original polymer [32]. At that time, Marvel and Inskeep [32] postulated that the observations were somehow related to aldehyde end groups in poly(vinyl acetate) and/or in poly(vinyl alcohol) which, in turn, could form acetals with hydroxyl groups of poly(vinyl alcohol). This concept was extended to the possibility that ketals also might form [33].

Wheeler and co-workers [34] insisted that there seems to be no way for acetals or acetal linkages to form during the polymerization of vinyl acetate. Unless such linkages were initially present, the hydrolytic degradation of the polymer cannot be attributed to hydrolysis of these *gem*-diethers. The claimed isolation of 2,4-dinitrophenylhydrazones from low molecular weight poly(vinyl alcohols) in Marvel and Inskeep [32] has been overlooked in subsequent discussions. It is conceivable that aldehydes had been used as chain-transfer agents in the preparation of the poly(vinyl acetate) under study. In subsequent stages of work these reagents might have reacted with 2,4-dinitrophenylhydrazine. It is also possible that residual ester groups had reacted with the substituted phenylhydrazine to form the corresponding phenylhydrazides. This would be difficult to differentiate from 2,4-dinitrophenylhydrazones, particularly in polymeric systems.

Flory and Leutner [35, 36] have hydrolyzed poly(vinyl acetate) to the corresponding poly(vinyl alcohol) and then treated the product with periodic acid and other reagents which attack 1,2-diols.

The expected head-to-tail sequence of monomer units leads only to 1,3-diols which would not be attacked. Occasional head-to-head, tail-to-tail sequences could be expected to be attacked by periodic acid with rupture of polymeric chains. From the changes in molecular weight as a result of such reaction, a measure of the number of 1,2-diol units that were present is obtained. From this, the percentage of head-to-head, tail-to-tail sequences in the chain can be calculated.

Commercial poly(vinyl acetate), as is well known, can be subjected either to acid or basic hydrolysis to form a completely hydrolized poly(vinyl alcohol). Upon reacetylation of this poly(vinyl alcohol), the regenerated poly(vinyl acetate) usually exhibits a reduction in molecular weight [34].

Poly(vinyl acetates) produced at low conversion or at low temperatures may be subjected to this process without significant changes in the molecular weight distribution. As indicated before, Marvel and Inskeep [32] had postulated that acetal (or ketal) linkages were ruptured during hydrolysis. The existence of acetal bonding in poly(vinyl acetate) is, however, difficult to visualize.

Table II illustrates the changes in viscosity of a series of polymers produced at high conversion of monomer with different average molecular weight distribution upon alkaline alcoholysis and reacetylation. To be noted first is that low molecular weight polymers are not degraded by the processing under consideration. This

TABLE II
CHANGE IN VISCOSITY OF POLY(VINYL ACETATE) UPON ALCOHOLYSIS AND REACETYLATION[a]

Initial intrinsic viscosity of PVAc samples	Approximately MW $(\times 10^{-6})^b$	Viscosity of PVAc after alcoholysis and reacetylation	Approximately MW $(\times 10^{-6})^b$
0.14[c]	0.021	0.14	0.021
0.49[c]	0.14	0.42	0.11
1.28[c]	0.585	0.84	0.31
1.95[c]	1.10	1.03	0.422
2.11[c]	1.24	1.08	0.453
3.29[c]	2.41	1.16	0.505
4.10[d]	3.35	1.13	0.485
4.30[e]	3.60	1.17	0.511

[a] From Ref. [34].
[b] Estimated by simplifying the equation of A. Berensniewicz *J. Polym. Sci.* **35**, 321 (1959) $[\eta] = 1.83 \times 10^{-4} M^{0.65}$, to read $[\eta] = 1.83 \times 10^{-4} \times 3M^2$.
[c] Suspension polymers.
[d] Bulk polymers produced from purified monomer by photopolymerization.
[e] Bulk polymers produced from purified monomers initial with benzoyl peroxide.

implies that low molecular weight polymers are not significantly branched. At high molecular weights it should be noted that there is a trend toward degradation to a "maximum" molecular weight average of approximately 500,000. Even a polymer of molecular weight 3,600,000 is degraded into the equivalent of approximately seven equal segments.

Studies of the effect of conversion on degradation indicate that the extent of degradability increases with increased conversions.

Polymerizations carried out to high conversion at $-30°C$ with highly purified monomers gave insoluble resins which, after alcoholysis and reacetylation, gave products with intrinsic viscosities as high as 3.65. This intrinsic viscosity is significantly higher than that obtained from polymers produced at 67°–70°C [34].

Hydrolysis under acidic conditions leads to degradation which is frequently not as extensive as that found with alkaline hydrolysis. In fact, there may be increases in the viscosity as a result of the reaction under acid conditions. This aspect of the observations has not been adequately explained.

Most of the chemical facts described have been attributed to branching of the polymer primarily through an ester bond.

The ester linkages presumably form by a chain-transfer reaction with the carbon–hydrogen bonds of the acetate grouping both in the polymer and in the monomer.

In Structure 1, 1, 2, and 3 indicate the hydrogens which may conceivably be involved in chain-transfer reactions leading to branching:

$$-\overset{3}{C}H_2-\overset{2}{C}H-$$
$$|$$
$$O$$
$$|$$
$$C=O$$
$$|$$
$$^1CH_3$$

Structure 1:

Branches at positions 2 or 3 would, of course, not be degradable under the conditions discussed here. A kinetic analysis indicated that branching at position 1, leading to a polymeric ester branch, was of primary importance in developing branched polymers [37].

The chain-transfer constant of the poly(vinyl acetate) is larger than that of most other monomers. Consequently branching is very significant in these polymerizations. Chain transfer takes place 40 times more frequently at position 1 than at position 2 [38]. Branching at position 3 is negligible. The number of branch points per polymer molecules may be computed as a function of the degree of conversion [39] as well as a function of temperature [21]. Emulsion polymerization of vinyl acetate at 5°C gave a polymer with very little branching. On the other hand, polymers, whether produced in emulsion, suspension, or solution at 65°C gave rise to approximately one branch for each polymer chain. At high temperatures and higher conversions, the molecular weight of poly(vinyl acetate) passes through a maximum. This effect is thought to be related to chain branching at approximately 0.15 branches per polymer molecule. The result is independent of the technique used in producing the polymer [40].

These observations are at variance with the data given in Table II where it would appear that (assuming that $[\eta]$ is a measure of single long chains) a polymer with MW 3.6×10^6 is made up, on the average, of approximately seven long branches of MW 5.1×10^5 joined together through position 1 (the acetoxy unit) in Structure 1.

To indicate that the nature of the branching in poly(vinyl acetate) is far from understood reference is made to the papers of Nozakura et al. [41–44]. These workers proposed to study branching in poly(vinyl alcohol). Obviously, if most of the branching in poly(vinyl acetate) occurs through the acetoxy linkage, the poly(vinyl alcohols) isolated from its hydrolysis should be essentially unbranched. Such a poly(vinyl alcohol) is reacetylated to give linear poly(vinyl acetate). In this work the possible low level of branches grafted onto position 2 of Structure 1, was ignored. The linear poly(vinyl acetate) was now only partially hydrolyzed. Through the resulting hydroxyl groups, the polymer was cross-linked with distilled commercial toluene diisocyanate (TDI) [42].

4. REACTIVITY RATIOS OF VINYL ESTERS

The reactivities of various vinyl esters in copolymerizations are generally very similar. The sampling of reactivity ratios given in Table III indicates this quite clearly. The somewhat unique behavior of vinyl benzoate is indicated in Table III.

The copolymerization of various substituted vinyl benzoates has been studied. It was found, first of all, that there was no polymerization with vinyl *p*-nitrobenzoate. A plot of the relative reaction rates

TABLE III
Reactivity Ratios of Selected Vinyl Esters

M_2	r_1	r_2	Ref.
Copolymers of various (M_2) with vinyl acetate (M_1)			
Vinyl formate	0.94	0.95	47
Vinyl propionate	0.98, 1.06	0.98, 0.76	47, 3[a]
Vinyl butyrate	1.00	0.97	47
Vinyl trimethylacetate ("Vinyl pivalate")	0.79	0.96	3
Vinyl 2-ethylhexanoate	1.19	1.90	3[b]
Vinyl neonanoate	0.93	0.90	3
Vinyl neodecanoate	0.99	0.92	3
Vinyl phenylacetate	0.96	0.92	47
Vinyl benzoate	0.70	1.13	47
Vinyl ethyl oxalate	0.30	3.00	48
Vinyl thioformate	0.05	5.50	46
Acrylic acid	0.02	20.64	c
Behenyl acrylate	0.021	1.76	49
Butyl acrylate	0.05, 0.05	5.89, 5.50	50[c]
	0.21	3.3	51
Crotonic acid	0.62	0.27	c
Ethyl acrylate	0.02	7.20	c
Ethylene	1.00	1.00	3
Maleic anhydride	0.01	0.01	c
Methyl methacrylate	0.03, 0.03	26.0, 22.21	47, 3
Phenyl acrylate	0.22	2.48	52
Vinyl chloride	0.60	1.40	c
Copolymers of vinyl esters (M_2) with methyl methacrylate (M_1)			
Vinyl formate	28.6	0.05	47
Vinyl acetate	26.0, 22.21	0.03, 0.03	47, 3
Vinyl propionate	24.0	0.03	47
Vinyl butyrate	25.0	0.03	47
Vinyl phenylacetate	26.4	0.03	47
Vinyl benzoate	20.3	0.07	47
Vinyl ethyl oxalate	6.0	0.10	48

[a] All the data from Ref. [3] are said to be calculated values (see footnote c).

[b] Private communication of E. L. Kitzmiller, Lehigh University to Union Carbide Chemicals and Plastics Co., South Charleston, WV, in Ref. [3].

[c] Calculated data from "Vynate, Vinyl Ester Monomers," Technical Bulletin F-60848 10/93-3M, Union Carbide Corporation, Danbury, CT, 1993, and D. Lee, *Am. Paint Coatings J.* Oct. 18, 1993, reprinted in Technical Bulletin F-60889 11/93-1M, Union Carbide Corp.

of vinyl benzoates with parasubstituents such as hydrogen, methoxy, methyl, chloro, bromo, and cyano groups against Hammett's σ-values gave a straight line. A small Hammett ρ-valued indicated that, in this case, a small polar effect operates on the vinyl group [45].

The radical copolymerization of vinyl thioacetate and vinyl thiobenzoate has also been investigated. Overall, the polymerization rate of vinyl thioacetate was smaller than that of vinyl acetate, but its reactivity in copolymers was larger. This has been attributed to the participation of a $d-\pi$ interaction of the sulfur groupings with the vinyl moiety [46].

5. CONFORMATION OF POLY(VINYL ESTERS)

The tacticity of the polymers of vinyl acetate and other vinyl esters has been studied for some time. Increasing stereoregularity of polymers was noted as esters of halogenated acids with increasing amounts of chlorine were polymerized. The sequence of increasing degree of stereoregularity was in the order: vinyl acetate < monochloroacetate < polychloroacetate < trifluoroacetate. Even in the most stereoregular poly(vinyl trifluoroacetate) ester, the sequence of regularly oriented units was relatively short [53]. Fordham and co-workers [53] observed that syndiotactic propagation may be preferred for free-radical polymerization in general. In other words, it was proposed that free-radical polymers may be significantly more regular in conformation than had previously been realized. They believed that the polymerization process was characterized by syndiotactic propagation, with frequent interruptions by isotactic propagation steps. It has been stated that at a conversion of 15%, the free-radical polymerization of vinyl acetate with AIBN and 2.4% triethylamine at 27°C produced an atactic poly(vinyl acetate) with a number-average molecule weight of 166,000. When chloroform was added to this initiator system, chain-transfer activity was sufficient to reduce the molecular weight of an atactic poly(vinyl acetate) to 1894 [54].

The vinyl esters of perfluorinated acids give rise to polymers which tend to be unstable to moisture. In fact, both the monomers and the polymers hydrolyze readily [55]. Interestingly enough, unlike the corresponding vinyl acetate system, poly(vinyl trifluoroacetate) is insoluble in its monomer. Substitution of fluorines for the hydrogens of the acetate portion of the monomer is thought to reduce chain transfer and branching. The poly(vinyl alcohols) derived from the hydrolysis of stretched poly(vinyl trifluoroacetate) were found to be highly ordered and birefringent [56].

Even in the presence of such aldehydes as acetaldehyde, propionaldehyde, butyraldehyde, and heptanol (all conventional chain-transfer agents for vinyl esters) in the temperature range from $-40°C$ to $60°C$, the free-radical polymerization of vinyl trifluoroacetate gives rise to a polymer in which syndiotactic and isotactic diads are found in up to 56% concentration [57].

Bulky substituents on the acid portion of vinyl ester evidently affect the stereoregularity of the polymer obtained either by free-radical or cationic mechanisms. To study this matter further, vinyl 1-adamantanecarboxylate, vinyl 1-adamentyl ether, and, as open-chain analogs of the adamentyl group, various vinyl trialkylacetates and vinyl tri-n-propylcarbinyl ethers were prepared and polymerized.

The admantanecarboxylate ester exhibited a syndiotactic propagation stage much like the trialkylacetates. Syndiotacticity of the polymers seems to increase as the molecular weight of the trialkylacetate moiety increases up to a limiting value of approximately 65% syndiotacticity for vinyl tri-n-butylacetate as is indicated in Table IV. Table IV also lists the general conditions for the preparation of these polymers by bulk polymerization in degassed, sealed ampoules. The final purification of the polymers was by reprecipitation from benzene solutions with methanol.

6. BULK POLYMERIZATION

The bulk polymerization of vinyl acetate is primarily of interest for laboratory studies, although a few large-scale procedures have been reported. Since the heat of polymerization is quite high (21 kcal/mole) and the boiling point of the monomer is relatively low (72.7°C) not only must the reaction temperature be monitored closely, but the reaction temperature must be kept low, unless pressure equipment is used. The low temperatures mean that the usual initiators of free-radical polymerization will act rather slowly. To further complicate bulk polymerizations, the polymerization process is strongly autocatalytic [17, 59].

TABLE IV
BULK POLYMERIZATION OF VINYL 1-ADAMANTANECARBOXYLATE AND
VINYL TRIALKYLACETATES[a,b]

Monomer	Initiator (moles)	Polymerization Temp (°C)	Time (hr)	Yield (%)	Softening point (°C)	Syndio-tacticity (%)
Vinyl 1-adamantane-carboxylate	0.2	60	5	89	220–225	53
Vinyl trimethylacetate	0.2	60	22	80	73–78	56
Vinyl tri-ethylacetate	0.1	45	24	96	80–83	61
Vinyl tri-n-propylacetate	0.05	60	6	87	51–55	62
Vinyl tri-n-butylacetate	0.1	60	24	84	68–70	64

[a] From Ref. [58].
[b] Polymerization conditions: The monomer concentration was 20 moles. Polymerization was conducted using 2,2'-azobisisobutyronitrile as the initiator in degassed sealed ampoules. Times and temperatures are given for each monomer in the body of the table.

Among the initiators which have been used in the bulk polymerization of vinyl acetate are dibenzoyl peroxide (BPO) [35, 60, 61], benzoyl steroyl peroxide, disteroyl peroxide, dialiphatic acyl peroxides in general, 2,2'-azobisisobutyronitrile [25, 61, 62], pinacols [63], dilauroyl peroxide (LPO), and difuroyl peroxide [53, 64].

Bulk polymerizations of vinyl acetate, on a laboratory scale, have been carried out in sealed tubes or ampoules, in dilatometers, and at atmospheric pressure at reflux.

In devising experimental procedures for the polymerization of vinyl esters, the elimination of oxygen is extremely important. Joshi [17] has shown that the bulk polymerization of vinyl acetate and vinyl propionate exhibited autoacceleration from the start and proceeded nearly to completion with only 2–4% unreacted monomer within 200 min in one case. When, however, the monomer had come in contact with air, inhibiting impurities developed and even after hours of heating, "dead-end polymerization" had taken place with 30–40% of unreacted monomer remaining.

We consider bulk polymerizations in sealed ampoules or even sealed heavy-walled tubes not merely unsafe, but dangerous. The procedure, in some respects is so trivial, that it is rarely described in any detail. Yet, so much polymer chemistry has been studied by sealed-tube polymerization that the procedure has to be described.

Procedure 23-1 is a composite of those described elsewhere [22, 41, 53, 65, 66]. Safety procedures will have to be designed to conform to OSHA regulations.

23-1. Sealed-Tube, Bulk Polymerization of Vinyl Acetate

In a heavy-walled Pyrex ampoule with a constriction near its upper opening is placed a dispersion of 10 ml of purified vinyl acetate and 0.0004 gm of 2,2'-azobisisobutyronitrile. The content is sparged with oxygen-free nitrogen or argon and attached to a high-vacuum line. The monomer is chilled with a dry ice–acetone mixture. The tube is evacuated on the high-vacuum line while the monomer remains frozen. By conventional techniques, the monomer is then degassed repeatedly. Finally the tube is sealed under reduced pressure.

After the seal has cooled, the tube is placed in a suitable, protective sleeve and slowly tumbled, end-over-end, in a constant temperature bath at $50 \pm 0.2°C$ with suitable safety precautions. Conversion, after 3 hr, is approximately 35.6%.

To isolate the product, the ampoule is cooled in dry ice–acetone, appropriately wrapped, and with suitable safety precautions, the ampoule is opened. The polymer may be dissolved in such solvents as acetone, benzene, or tetrahydrofuran (THF) and precipitated with petroleum ether or with redistilled hexane. The polymer may also be isolated from benzene solution by freeze-drying.

Bulk polymerizations have been carried out in a conventional reflux apparatus. While, in principle, this procedure permits the preparation of larger quantities of polymer than is possible in sealed tubes, the control of the reaction temperature presents problems. The polymerization is autoaccelerating. As the viscosity of the medium increases and the molecular weight increases, heat transfer from the interior of the reacting mass becomes increasingly difficult. The heat of polymerization is quite high so that great care is required. Naturally, a means of removing the polymer mass from the equipment also needs to be provided. A pneumatic jack and temperature controller will be useful for bulk processes.

7. SOLUTION POLYMERIZATION

The polymerization of vinyl acetate in solution may be carried out to produce lacquers, chewing gum bases, and adhesives. Usually the polymers are used directly in the solvents in which they are formed. A few applications may call for the recovery of the solid polymer from solution. This may be accomplished by addition of a nonsolvent of the polymer or, where applicable, by removal of the solvent by steam distillation [67].

Most solvents act as chain-transfer agents in the polymerization of vinyl acetate. The monomer itself may act as a chain-transfer agent for its own radicals. *tert*-Butyl alcohol is exceptional in that it is one of the few common solvents with minimal, if any, chain-transfer activity [68, 69].

Of the common solvents, *tert*-butyl alcohol because of its very low chain-transfer constant, may be used to produce polymers of relatively high molecular weight. If we concede that single-point measurements of specific viscosity and inherent viscosity may be considered indications of the general trend of molecular weights, then the effect of various solvents on the molecular weights of the poly(vinyl acetate) produced may be seen in Table V.

Procedure 23-2 outlines the method for polymerizing vinyl acetate in *tert*-butyl alcohol. Particular attention is directed to the method for removing excess monomer and solvent.

23-2. Polymerization of Vinyl Acetate in *tert*-Butyl Alcohol [68]

To equipment involving a 500 ml three-necked flask fitted with a propeller-type agitator, thermometer, reflux condenser, and inlet

TABLE V
THE EFFECT OF SOLVENTS ON THE POLYMERIZATION OF VINYL ACETATE[a,b]

Diluent	Yield of polymer (%)	Specific viscosity	Inherent viscosity (dl/gm)
(None)	92.4	0.210	1.89
Isopropyl alcohol	75.4	0.023	0.22
n-Propyl alcohol	50.0	0.031	0.29
Acetic acid	34.0	0.047	0.44
Ethyl acetate	73.2	0.080	0.75
tert-Butyl alcohol	84.2	0.186	1.69

[a] From Ref. [68].
[b] Polymerization conditions: Polymerization was conducted using 50 gm of vinyl acetate, 30 ml of diluent, and 0.05 gm of dibenzoyl peroxides. The solution was heated for 22 hr at reflux.

and outlet tubes for purified nitrogen that are fitted with a bubble counter is added 50 gm of freshly distilled vinyl acetate dissolved in 30 ml of *tert*-butyl alcohol. To this solution is added 0.05 gm of dibenzoyl peroxide. The mixture is heated under nitrogen on a steam bath for 22 hr. The warm viscous solution is diluted with 200 ml of acetone. The solution is transferred to a 3 liter flask containing 1 liter of hot, distilled water to precipitate the polymer. The equipment is set up for steam distillation, steam is passed through the suspension to distill out unreacted vinyl acetate and *tert*-butyl alcohol.

After the steam distillation is completed, the polymer is filtered off and dried to constant weight at 80°C and 8 mm Hg pressure (yield, 42.1 gm or 84.2%; specific viscosity, 0.186; inherent viscosity in benzene, 1.69 dl/gm).

By using greater levels of *tert*-butyl alcohol, polymers with somewhat lower molecular weights are produced. As indicated in Table V, the chain-transfer activity of this solvent is substantially less than that of many other common solvents. This solvent has also been suggested for continuous polymerization or copolymerization processes [69].

Usually solution polymerizations are initiated, as shown above, with a single solvent-soluble initiator such as dibenzoyl peroxide. In a Romanian publication, a redox initiator consisting of *tert*-butyl-2-ethylperhexanoate and ascorbic acid was described as being particularly effective in the polymerization of vinyl acetate in a methanol solution. Polymerizations in the temperature range of 20–40°C were readily carried out. Since these reaction temperatures are substantially below the boiling point of the monomer, such redox initiation systems should be explored further [70].

8. SUSPENSION POLYMERIZATION

In general, we draw a sharp distinction between suspension and emulsion polymerization processes. This distinction is quite readily apparent in the case of monomers which are quite insoluble in water, such as styrene. In that case, by use of monomer-soluble initiators and a variety of suspending agents, the suspension-polymerization process leads to the formation of spherical particles which can be separated by filtration.

When water-soluble initiators and surface-active agents are used, relatively stable lattices are formed from which the polymer cannot be separated by filtration. In the case of vinyl acetate, the distinctions are more blurred.

Between the true suspension and the true emulsion polymerization, we find, according to Bartl [4], the processes for formation of reasonably stable dispersion of fine particles of poly(vinyl acetate) using reagents which are normally associated with suspension polymerization. The product is described as "creme-like." The well-known white, poly(vinyl acetate), household adhesives may very well be examples of these creamy dispersions. The true lattices are characterized by low viscosities and particles of 0.005-1 μm diameter. The creme-like dispersions exhibit higher viscosities and particle diameters of 0.5-15 μm.

Probably most industrial homopolymerizations of vinyl acetate are carried out by suspension processes. Surprisingly little has been published about the suspension polymerization of vinyl acetate, despite the fact that Lindemann [1] cites nearly 1,100 references on the subject of vinyl acetate and approximately 200 references on the higher vinyl esters.

The polymerization of vinyl acetate is best carried out in a pH range in which the hydrolysis of the monomer is minimized. Since this hydrolysis leads to the formation of acetaldehyde, a notorious chain-transfer agent, careful control of the pH is important. The pH range of 4-5 is considered optimum for minimizing vinyl ester hydrolysis. Formic acid, at a level of 0.15-0.25% of the monomer has been suggested for this purpose [4]. In connection with this it should be kept in mind that formic acid does have an aldehydic structure and may, therefore act as a chain-transfer agent. It is also a potent reducing agent and may create a redox system with BPO or other oxidizing initiators.

Typical suspending agents for the vinyl acetate polymerization are poly(vinyl alcohol) [particularly a grade represented as approximately 88% hydrolyzed poly(vinyl acetate)], gum arabic, hydroxyethyl cellulose, methyl cellulose, starches, sodium polyacrylate or sodium polymethacrylate, gelatin, and an equimolar copolymer of styrene and maleic anhydride neutralized with either sodium hydroxide or aqueous ammonia. Water-insoluble dispersing agents or high concentrations of electrolyte to reduce water solubility are not usually used in the suspension polymerization of vinyl acetate.

The usual initiators are monomer-soluble ones such as dibenzoyl peroxide, lauroyl peroxide, and di-*o*-toluyl peroxide. Hydrogen peroxide and a few other water-soluble initiators usually associated with emulsion polymerizations have also been used. The molecular weight of the polymer may be controlled by variations in the concentration of the initiator. This effect is illustrated in Procedure 23-3. It is interesting to note that this procedure goes back to FIAT Final Report 1102, i.e., a report compiled toward the end of World War II [4]. It was still used in 1974 according to Bravar *et al.* [71]. We have adapted these procedures to a laboratory scale.

23-3. Suspension Polymerization of Vinyl Acetate (Control of Molecular Weight by Variation in Initiator Level) (Based on Bartl [4] and Bravar *et al.* [71])

(a) Preparation of low molecular weight poly(vinyl acetate). In a 3 liter reaction kettle fitted with a mechanical stirrer, reflux condenser, thermometer, and an addition funnel, 800 ml of distilled water and 0.8 gm of the sodium salt of an equimolar copolymer of styrene and maleic anhydride (German trade name Styromal) are heated to 80°C with agitation. Meanwhile a solution of 600 gm of vinyl acetate, 5.4 gm of dibenzoyl peroxide, and 3 gm of ethyl acetate is prepared.

To the warm aqueous suspending agent is added 100 gm of the vinyl acetate solution. The stirred mixture is brought up to 80°C by external heating. Once the polymerization has started, heating and cooling is applied as required while the remainder of the monomer solution is added over a 5 hr period. After the addition has been completed, heating is continued for an additional 3 hr period. The residual monomer is removed by steam distillation with agitation. The aqueous dispersion is cooled with agitation to 4°C. The polymer beads are filtered off or centrifuged and washed repeatedly with water at 5°C to remove the suspending agent. The polymer is then dried under reduced pressure at 30°C. The dry product is glass clear. The product is reported to have MW 110,000.

(b) Preparation of an intermediate molecular weight poly(vinyl acetate). For this preparation, the procedure used is the same as that given in the preceding section except that the reactants used are 800 ml of distilled water and 0.8 gm of Styromal (sodium salt)

to which is added by the described gradual addition technique a solution of 600 gm of vinyl acetate and 1.2 gm of dibenzoyl peroxide.

The final product has MW on the order of 1,000,000.

(c) Preparation of a high molecular weight poly(vinyl acetate). For this preparation, the procedure used is the same as that given in Section (a) except that the reactants used are 800 ml of distilled water and 1.2 gm of Styromal (sodium salt) to which is added by the described gradual addition technique a solution of 600 gm of vinyl acetate and 0.18 gm of di-*o*-toluyl peroxide.

The final product has MW on the order of 1,500,000.

A patented procedure for the suspension polymerization of vinyl acetate which is claimed to produce a nonsticky bead polymer uses an aqueous phase consisting of 537 gm of distilled water, 0.25 gm gum tragacanth, and 0.10 gm sodium dioctylsulfosuccinate (Aerosol OT). The monomer charged consists of 690 gm of vinyl acetate and 0.69 gm of dibenzoyl peroxide.

9. EMULSION POLYMERIZATION

From the industrial standpoint, the suspension polymerization of vinyl acetate is of primary interest for the production of poly(vinyl acetate) homopolymer beads. Most of these beads are converted to poly(vinyl alcohol) with a variety of degrees of hydrolysis and in a number of different molecular weight ranges. On a laboratory scale, suspension copolymerizations of vinyl acetate with other monomers may have an advantage in terms of ease of handling and ease of developing a suitable suspension medium. Industrially emulsion polymerizations, both homo- and copolymerizations, are of great importance, particularly in the development of adhesives, paints, paper coatings, and textile finishes.

The production of vinyl acetate monomer by the current top four U.S. producers (Höchst-Celanese, Union Carbide, Du Pont, and Quantum Chemical) in 1989 was about 1.25×10^9 kg [8], compared to 1.14×10^9 kg reported for 1980 [7] and 0.89×10^9 kg for 1977 [72]. Since there are major producers of poly(vinyl acetate) not only in the U.S. but also in Japan, the United Kingdom, Germany, Switzerland, Canada, France, and Italy, one may assume that the total annual production of the monomer is well above the figure given here for the U.S. production.

Although more than one-third of all the monomer is used to produce latices in the form of paints and adhesives, published information on the emulsion polymerization of vinyl acetate is limited. References [1, 4, 7, 8] are general references. Bacon [73] reviews the redox initiation of the polymerization. Shapiro [74] deals with the applications of vinyl acetate to the paper industry.

The emulsion polymerization of vinyl acetate may be unique among polymerization processes in that true latices have been formed with anionic surfactants, cationic surfactants, nonionic surfactants, or protective colloids, and with combinations of two or more such reagents, as well as without any added emulsifier.

During the emulsion polymerization of vinyl acetate, unlike the case of the styrene polymerization, emulsion particles form up to a conversion of 80%. The reaction also appears to be dependent on the rate of agitation; the more vigorous the stirring, the slower the rate. It is also reported that the effect of the initiator concentration (i.e., the concentration of a persulfate) is complicated by the formation of free sulfuric acid during the reaction. This leads to the hydrolysis of some of the monomer to acetaldehyde which, aside from its chain-transfer activity, also retards the rate [75]. These observations again point up the importance of pH control during the polymerization of vinyl acetate—a matter already mentioned in connection with suspension polymerizations. As a matter of fact, the rate of emulsion polymerization of vinyl acetate is said to be at a maximum at a pH of 7 [76]. This is higher than the range of pH 4–5 which is the one most desirable from the standpoint of minimal hydrolysis of vinyl esters [4].

An additional factor that may not have been considered adequately in the theoretical treatment of the emulsion polymerization of vinyl acetate arises from the work of Dunn and Taylor [77]. These researchers noted that in their dilatometric study of the emulsion polymerizations, the contraction of the monomer in an aqueous system was only $15.7 \pm 0.4\%$, whereas the bulk polymerization contraction was reported to be 26–28% [78]. The difference was attributed to the solvation of the monomer in the aqueous medium. If this is indeed so, it is not inconceivable that under the conditions of both suspension and emulsion processes, not the monomer vinyl acetate, but new monomers, of variable composition and conceivably of distinctly different chemical properties, are involved: vinyl acetate hydrates.

In the preparation of emulsion polymers, particularly when copolymer systems are involved, several methods of adding the monomer to the reacting system are available.

One may add all of the monomer at once to the aqueous phase. Alternatively, the monomers may be added gradually stepwise. Finally, the monomers may be added gradually and continuously. The first two methods lead to more or less heterogeneous copolymers while the continuous addition method affords homogeneous copolymers. The best adhesive properties are achieved from homogeneous copolymer systems; other properties may vary considerably with the degree of heterogeneity of the polymer [79]. There are, of course, procedures which also call for the gradual addition of initiator solutions, additional surfactant, plasticizer, etc.

The manipulation of various emulsion polymerization techniques assists in developing poly(vinyl acetate) latices with specific characteristics needed for certain end-uses. Some of these are summarized in Table VI.

Most emulsion polymerizations are, of course, carried out in conventional stirred reactors. There are procedures in which all of the monomer is charged at once to the reactor; procedures wherein the monomer is gradually added, possibly with gradual addition of surfactant solutions, initiator solutions, and other modifying agents; procedures using redox initiation or thermal initiation; and procedures using preemulsified monomers.

Procedure 23-4 illustrates a simple emulsion polymerization procedure. This system makes use of potassium persulfate as the initiator and sodium lauryl sulfate as the surfactant for the preparation of a vinyl acetate homopolymer latex.

23-4. Emulsion Polymerization of Vinyl Acetate–Potassium Persulfate–Sodium Lauryl Sulfate System [80]

To a 500 ml resin kettle with a bottom stopcock, equipped with a pressure-equalizing addition funnel, reflux condenser (topped with a pressure regulator permitting exhausting of nitrogen but preventing air from entering the system), a thermometer, a four-bladed paddle, 3.75 cm in diameter with blades set at 90° to each other in the reaction chamber with inside diameter of 7.5 cm, four baffle plates

TABLE VI
EFFECT OF POLYMERIZATION TECHNIQUE MODIFICATION ON LATEX CHARACTERISTICS

Technique modification	Effect
Increasing the concentration of surfactant and protective colloid	Latex viscosity increases; particle size decreases
Decreasing the concentration of surfactant and protective colloid	Particle size increases
Delayed addition of surfactant and protective colloid	Increases particle size
Delayed addition of monomer	Reduces particle size
Increasing the initial monomer charge 15–20%	Increases MW
Reducing the rate of delayed monomer addition	Reduces particle size
Use of water-soluble initiators	Reduces particle size
Use of monomer-soluble initiators	Increases particle size
Increasing initiator concentration	Reduces particle size: effect on MW is minimal
Decreasing initiator concentration	Increases particle size
Delayed addition of initiator	Increases particle size
Use of a redox initiator	May increase MW
Increasing the temperature of polymerization	Increases latex viscosity: reduces particle size
Decreasing the temperature of polymerization	Increases particle size: increases MW
Increasing the potential solids concentration of the latex	Increases latex viscosity
Decreasing or eliminating materials with chain-transfer activity	Increases MW

approximately 0.75 cm wide, and a nitrogen inlet, were added 250 ml of distilled water, 50 gm of vinyl acetate, and 0.25 gm of sodium lauryl sulfate. The mixture was freed of oxygen by bubbling nitrogen through it for at least 0.5 hr. Then 0.3125 gm of potassium persulfate was added (possibly in aqueous solution which had been deoxygenated separately). The agitator was operated at 400 rpm and the reaction is maintained at 50 ± 0.5°C. Within 40 min, nearly 100% conversion was observed.

The emulsion polymerization of vinyl esters of the high carboxylic acids is somewhat difficult since stable emulsification before and during polymerization is difficult to achieve with the common soaps of the alkylaryl sulfonate salts. Best results are said to be achieved when a mixture of a nonionic surfactant such as esters of anhydrosorbitol or anhydromannitol (e.g., Span 20) and a branched-chain alcohol sulfate [e.g., Tergitol Paste 4 (50%)] are used. These emulsifiers do not produce emulsion polymers with vinyl acetate but do so with vinyl palmitate. The usual procedure for polymerization consists of weighing monomer and emulsifiers into a flask through which oxygen-free nitrogen is passed. With swirling and warming, a uniform mixture is produced which is transferred to a mechanical blender containing oxygen-free distilled water. In the blender, during agitation, nitrogen is continuously passed through. The resultant emulsion is then polymerized in conventional equipment for 7 hr at 55°–75°C. A typical recipe consists of the following:

Reagant	Amount
Vinyl palmitate	0.20 moles
Span 20	2.8 gm
Tergitol Paste 4 (50%)	5.7 gm
Water, distilled	113 gm
Potassium persulfate	0.27 gm (0.001 moles)

Within 7 hr at 55°–75°C, a conversion greater than 80% was achieved. To isolate the polymer, the latex may be coagulated by pouring it into a warm aqueous sodium chloride solution. The polymer may be purified by dissolving it in benzene and then, at −5°C pouring the solution while swirling by hand into acetone. The poly(vinyl palmitate) could be isolated as discrete particles [81].

The monomer vinyl pivalate polymerizes readily. Unlike most common monomers, this monomer yields polymers of high molecular weight by the suspension procedures and modest molecular weights in an emulsion polymerization. This has been attributed to its lower capability of diffusing through the aqueous system as compared to vinyl acetate. This monomer is quite resistant to saponification and exhibits little chain-transfer activity [82].

Procedure 23-5 is an adaptation of a ter-polymerization that starts with the formation of a seed latex followed by the gradual addition of both a monomer composition and an initiator solution at separate rates. The resulting latex has a high percentage of nonvolatiles. It is said to be suitable for formulating good emulsion paints. In connection with this preparation, care must be taken that the initiating ammonium persulfate is indeed active.

23-5. Emulsion Ter-polymerization of Vinyl Acetate, Butyl Acrylate, and Vinyl Neodecanoate (Seeded Process with Gradual Monomer and Initiator Additions) [83]

To a 1 liter resin kettle equipped with a reflux condenser, an explosion-proof stirrer, a thermometer and temperature controller, and a nitrogen inlet is charged 204.00 gm of deionized water. Then 6.00 gm of Cellosize hydroxyethyl cellulose WP-300, 3.00 gm of Tergitol NP-40, 3.90 gm of Tergitol NP-15 (two nonionic surfactants), 3.3 gm of Siponate DS-4 (an anionic surfactant), and 0.6 gm of ammonium bicarbonate are added. The stirred mixture is blanketed with nitrogen and warmed to 55°C. This temperature is maintained for 20 min. A mixture of 18.00 gm of vinyl acetate, 4.50 gm of butyl acrylate, and 7.5 gm of vinyl neodecanoate is added, followed by 0.24 gm of ammonium persulfate. The reaction mixture is heated to 75°C and maintained at that temperature for 15 min to form the seed latex.

The reaction temperature is then raised to 78°C and the gradual addition of monomers and initiator solution is begun. The monomer solution, consisting of 162 gm of vinyl acetate, 41.4 gm of butyl acrylate, and 66.6 gm of vinyl neodecanoate, is added over a 2 hr period. The initiator solution of 0.60 gm of ammonium persulfate dissolved in 60.00 gm of deionized water is added over a period of 2.5 hr. All the while, the reaction temperature is maintained at 78°.

After the last of the initiator solution has been added, heating and good stirring is continued for another hour. The latex is cooled and filtered through a 200 mesh stainless steel screen. The percent nonvolatiles of the latex is 53.2%. Coagulum is 0.01%.

Vinyl acetate is fairly water soluble and somewhat deficient in hydrolytic stability. The incorporation of a vinyl ester of the higher branched carboxylic acids into a copolymer system improves the resistance of the product to hydrolysis. Thus the latex formed in Procedure 23-5 has been used in formulating exterior paints. Copolymers of vinyl acetate with increasing concentrations of vinyl 2-ethylhexanoate (VEH) were prepared by a procedure similar to that of Procedure 23-5, except that the initiator system consisted of *tert*-butyl hydroperoxide and sodium formaldehyde sulfoxylate, and the buffer was sodium acetate. As the level of VEH increased, the hydrolytic stability of the copolymer increased significantly [84].

10. NONAQUEOUS DISPERSION POLYMERIZATION

The preparation of polymers as dispersions in nonaqueous systems has not been discussed very extensively. Even so, an interesting book has been published [85].

Stable dispersions in organic liquids are usually formed by using graft copolymers to stabilize the system. According to one patent, the copolymer "emulsifier" is selected so that its polymer chains contain groups of differing polarities; some of which may be solvated by the organic solvent and other groups which may become associated with the particles of the dispersed polymer [86]. These stable dispersions have found application in baked-on finishes.

In Procedure 23-6, a remarkably high solids dispersion of poly(vinyl acetate) in cyclohexane is prepared. Since this procedure has been patented, it is given here only to illustrate the techniques used in preparing nonaqueous polymer dispersions.

23-6. Preparation of Poly(vinyl acetate) in Nonaqueous Dispersion [87]

In a resin kettle equipped with a mechanical stirrer, reflux condenser, thermometer, and provisions for passing oxygen-free nitrogen through the system, with agitation, under a nitrogen atmosphere, 240 gm of vinyl acetate, 160 gm of cyclohexane, and 1.4007 gm of an ethylene–vinyl acetate copolymer (DQDA 3267, a copolymer containing 28 wt % vinyl acetate and having a melt index of 23.8 dg/min) are heated to 70°C for approximately 1 hr to bring about complete solution. Then 0.1514 gm of dibenzoyl peroxide is rapidly added. After stirring and heating for about 1.5 hr a milky dispersion begins to form. The polymerization is continued for 22 hr. The product is a dispersion with solids content of 58% at 97% conversion of the monomer to polymer. The poly(vinyl acetate) exhibited an inherent viscosity (0.2% solution in cyclohexane at 30°C) of 45. The particle size of the polymer is between 0.3 and 5 μm.

This dispersion can be used to produce a dispersion of poly(vinyl alcohol) [albeit only about 75% hydrolyzed poly(vinyl acetate)] by hydrolyzing it with a solution of sodium methylate in methanol at 30°C.

Dispersions of poly(vinyl acetate) in n-alkanes have been prepared using a diblock copolymer of poly(styrene-b-[ethylene-copropylene)] as the stabilizer for the colloidal PVAc. The dispersing agent contained 38.5% styrene and was used at a concentration of 1–5 wt %. The monomer concentration ranged from 10 to 30 wt % in the various experiments. A typical initiator was 2,2′-azobisisobutyronitrile. The particle diameters were in the range of 0.10 to 0.31 μm [88].

11. RADIATION-INITIATED POLYMERIZATION

Many details of the complex steps in the formation of polymers of high molecular weight beginning with the initiation step have been studied by UV-radiation-initiated polymerization.

By a technique which interposed a rotating sector between a source of UV radiation and a dilatometer bearing a monomer, variations in the rotational speed of the sector and size of the opening, controlled bursts of radiation strike the monomer and induce polymerization. From the frequency of exposure and the effect on the polymerization many of the kinetic constants were evaluated. In this connection it should be noted that vinyl acetate exhibits virtually no absorption of UV radiation at 290–300 nm. On the other hand, acetaldehyde has an extinction coefficient of 14 at 290 nm and an extinction coefficient of 15 at 300 nm. Therefore acetaldehyde can act as a photosensitizer for the polymerization of vinyl acetate at wavelengths above 299.8 nm [89]. At 366 nm, 2,2′-azobisisobutyronitrile has been used as a sensitizer [23, 90]. Azobicyclohexane carbonitrile is a UV sensitizer suitable for use with a 124 watt mercury arc at 25°C which does not produce a dark reaction in rotating-sector experiments. Its absorption peak is at 350 nm with an extinction coefficient of 16 [91].

The UV-initiated polymerization of vinyl acetate, sensitized with various quantities of AIBN at -19°C, gives polymeric products which, upon saponification and reacetylation, produce polymers with substantially the same molecular weight as that initially observed. These observations contribute to the postulate that at low temperatures, polymerizations proceed essentially with the formation of branched chains [20].

A composition of vinyl acetate, ethylene dimethacrylate, benzil (a UV sensitizer), and finely ground vinyl acetate-ethylene dimethacrylate popcorn copolymer on exposure to radiation at 365 nm gave rise to a proliferating polymerization which continued after the source of radiation had been turned off. It has been postulated that in this case most of the growing radicals are formed by chain scission during polymerization [92].

Vinyl acetate, frozen to a glassy state or a crystallized form, has been subjected to solid-state polymerization with UV radiation. The rate of polymerization was found to vary with the physical state [93]. On the other hand at $-195°C$, vinyl acetate did not polymerize when exposed to a dosage rate of 7.8×10^5 rad (γ radiation) [93a].

The "living" radical polymerization of vinyl acetate and subsequent use of the "living" polymer to form block copolymers makes use of the initiating system consisting of triisobutyl aluminum, 2,2'dipyridyl, and the 2,2,6,6-tetramethyl-1-piperidinyloxy free radical ("TEMPO"). The molar ratio of the trialkyl aluminum to Tempo was 3 to 2; that of the trialky aluminum to dipyridyl was 1 to 1. At room temperature, 90% conversion of the monomer was achieved in 1 day. The molecular weight of the product, M_n, was 50,000, with polydispersity ranging from 1.25 to 1.5. As more monomer was added to the system, the molecular weight increased. When other monomers, such as methyl methacrylate or styrene, were added, block copolymers formed. This was considered a characteristic of "living" polymerization. The authors claim that this work represented the first cases of block-copolymer formation with poly(vinyl acetate) as one of the blocks [94].

The use of trialkylboron and related initiators is discussed in Refs. [95–97].

The polymerization of vinyl esters with trialkylaluminum and organic peroxide was patented in 1974 [98].

Zinc chloride has been used in a variety of polymerization systems. References to its use and to charge transfer complex polymerizations are Nikolayev *et al.* [23], Imoto *et al.* [65], Serniuk and Thomas [99], and Seymour *et al.* [100].

REFERENCES

1. M. K. Lindemann, *Encycl. Polym. Sci. Technol.* **15**, 531 (1971).
2. M. K. Lindemann, *in* "Vinyl and Diene Monomers" (E. C. Leonard, ed.), Part 1, pp. 329ff, Wiley (Interscience), New York, 1970.
3. O. W. Smith, M. J. Collins, P. S. Martin, and D. R. Bassett, *Progr. Org. Coat.* **22**, 19 (1993).

4. H. Bartl, in "Houben-Weyl Methoden der organischen Chemie" (E. Müller, ed.), 4th ed., Vol. 14, Part 1, pp. 905ff, Thieme, Stuttgart, 1961.
5. E. Tromsdorff and C. E. Schildknecht, in "Polymer Processes" (C. E. Schildknecht, ed.), Wiley (Interscience), New York, 1956.
6. M. S. E. El-Aasser and J. W. Vanderhoff (eds.), "Emulsion Polymerization of Vinyl Acetate," Applied Science Publishers, Englewood Cliffs, New Jersey, 1981.
7. W. Daniels, in "Kirk-Othmer, Encyclopedia of Chemical Technology," 3rd ed., Vol. 23, pp. 817ff, Wiley, New York, 1983.
8. W. A. Daniels, in "Concise Encyclopedia of Polymer Science and Engineering" (J. I. Kroschwitz, exec. ed.), pp. 1264ff, Wiley, New York, 1990.
9. E. Rohde, *Rubber World* **208**(3), 36, 58 (1993).
10. D. L. Cincera, in "Kirk-Othmer, Encyclopedia of Chemical Technology," 3rd ed., Vol. 23, pp. 848ff, Wiley, New York, 1983.
11. F. L. Marten, in "Concise Encyclopedia of Polymer Science and Engineering" (J. I. Kroschwitz, exec. ed.), pp. 1264ff, Wiley, New York, 1990.
12. P. Layman, *Chem. Eng. News* (March 22, 1993), 8.
13. E. T. Wise and S. G. Weber, *Am. Chem. Soc. Poly. Prep.* **34**(1), 215 (1993).
14. M. K. Lindemann, in "Kinetics and Mechanisms of Polymerization" (G. E. Ham, ed.), Vol. 1, Part 1, Chapter 4, Dekker, New York, 1967.
15. A. Vranken and G. Smets, *Makromol. Chem.* **30**, 197 (1959).
16. K. Nozaki and P. D. Bartlett, *J. Am. Chem. Soc.* **68**, 2377 (1946).
17. R. M. Joshi, *J. Polym. Sci.* **56**, 313 (1962).
18. G. M. Burnett, M. H. George, and H. W. Melville, *J. Polym. Sci.* **16**, 31 (1955).
19. W. R. Sorensen and T. Campbell, "Preparative Methods of Polymer Chemistry," pp. 171ff, Wiley (Interscience), New York, 1961.
20. L. M. Hobbs, S. C. Kothari, V. C. Long, and G. C. Sutaria, *J. Polym. Sci.* **22**, 123 (1956); L. M. Hobbs and V. C. Long, *Polymer* **4**, 479 (1963).
21. G. V. Schulz and L. Roberts-Nowakowska, *Makromol. Chem.* **80**, 36 (1964).
22. W. W. Graessley and H. M. Mittelhauser, *J. Polym. Sci., Part A-2* **5**, 431 (1967).
23. A. F. Nikolayev, M. E. Rozenberg, V. A. Kuznetsova, T. V. Kreitser, and G. S. Popova, *Vysokomol. Soedin., Ser. A* **15**(7), 1440 (1973); *Polym. Sci. USSR (Engl. Transl.)* **15**, 1610 (1973).
24. H. Gunesch and I. A. Schneider, *Makromol. Chem.* **132**, 259 (1970).
25. W. H. Stockmeyer and L. H. Peebles, *J. Am. Chem. Soc.* **75**, 2278 (1953).
26. J. T. Clarke, R. O. Howard, and W. H. Stockmeyer, *Makromol. Chem.* **44 / 46**, 427 (1961).
27. M. Matsumoto and M. Maeda, *J. Polym. Sci.* **17**, 435 (1955).
28. M. Litt and V. Stannett, *Makromol. Chem.* **37**, 19 (1960).
29. M. Matsumoto and M. Maeda, *J. Polym. Sci.* **17**, 438 (1955).
30. J. C. Bevington and M. Johnson, *Eur. Polym. J.* **4**, 373 (1968).
31. J. C. Bevington and M. Johnson, *Eur. Polym. J.* **4**, 669 (1968).
32. C. S. Marvel and G. E. Inskeep, *J. Am. Chem. Soc.* **65**, 1710 (1943).
33. J. T. Clarke and E. R. Blout, *J. Polym. Sci.* **1**, 419 (1946).
34. O. L. Wheeler, S. L. Ernst and R. H. Crozier, *J. Polym. Sci.* **8**, 409 (1952).
35. P. J. Flory and F. S. Leutner, *J. Polym. Sci.* **3**, 880 (1948).

36. P. J. Flory and F. S. Leutner, *J. Polym. Sci.* **5**, 267 (1950) (errata to Flory and Leutner [35]).
37. O. L. Wheeler, E. Levin, and R. N. Crozier, *J. Polym. Sci.* **9**, 157 (1952).
38. S. Imoto, J. Ukida, and T. Kominami, *Kobunshi Kagaku* **14**, 101 (1957).
39. D. J. Stein, *Makromol. Chem.* **76**, 170 (1964).
40. S. G. Ragova, S. S. Mnatsakanov, E. S. Shul'gina, M. E. Rozenberg, A. F. Smirnov, and S. Ya. Frenkel, *Tr. Probl. Lab., Leningr. Inst. Tekst. Legk. Prom.* No. 13, p. 526 (1971); *Chem. Abstr.* **78**, 30387j (1973).
41. S.-I. Nozakura, Y. Morishima, and S. Murahashi, *J. Polym. Sci., Part A-1* **10**, 2853 (1972).
42. S.-I. Nozakura, Y. Morishima, and S. Murahashi, *J. Polym. Sci., Part A-1* **10**, 2767 (1972).
43. S.-I. Nozakura, Y. Morishima, and S. Murahashi, *J. Polym. Sci., Part A-1* **10**, 2781 (1972).
44. S.-I. Nozakura, Y. Morishima, and S. Murahashi, *J. Polym. Sci., Part A-1* **10**, 2867 (1972).
45. M. Kinoshita, T. Irie, and M. Imoto, *Makromol. Chem.* **110**, 47 (1967).
46. M. Kinoshita, T. Irie, and M. Imoto, *Kogyo Kagaku Zasshi* **72**, 1210; English abstract see **72** (No. 5), A63 (1969).
47. J. C. Bevington and M. Johnson, *Eur. Polym. J.* **4**, 669 (1968).
48. N. Kawabata, T. Tsuruta, and J. Furukawa, *Makromol. Chem.* **48**, 106 (1961).
49. B. Subrahmanyam, S. D. Baruah, M. Rahman, J. N. Baruah, and N. N. Dass, *J. Polym. Sci., Part A: Polym. Chem.* **30**, 2273 (1992).
50. C. Pichot, M. F. Llauro, and Q. T. Pham, *J. Polym. Sci., Polym. Chem. Ed.* **19**, 2619 (1981).
51. M. A. Abd El-Ghaffar, A. S. Badran, and S. M. M. Shendy, *J. Elastomers Plast.* **24**(3), 192 (1992).
52. C. S. Ha, Y. K. Kim, D. P. Kang, D. W. Kim, S. D. Seul, and W. J. Cho, *Pollimo* **16**(6), 646 (1992); *Chem. Abstr.* **118**, 102555g (1993).
53. J. W. L. Fordham, G. H. McCain, and L. E. Alexander, *J. Polym. Sci.* **39**, 335 (1959).
54. W. R. Brown and G. S. Park, *Am. Chem. Soc., Div. Org. Coat. Plast. Chem., Pap.* **34**(1), 505 (1974).
55. T. S. Reid, D. W. Codding, and F. A. Bovey, *J. Polym. Sci.* **18**, 417 (1955).
56. H. C. Haas, E. S. Emerson, and N. W. Schuler, *J. Polym. Sci.* **22**, 291 (1956).
57. S. Matsuzawa, K. Yamaura, and H. Noguchi, *Makromol. Chem.* **168**, 27 (1973).
58. S.-I. Nozakura, T. Okamoto, K. Toyora, and S. Murahashi, *J. Polym. Sci., Part A-1* **11**, 1043 (1973).
59. H. B. Lee and D. T. Turner, *Polym. Prepr., Am. Chem. Soc., Div. Polym. Chem.* **18**(2) 539 (1977).
60. A. Voss and W. Heuer, German Patent 666,866 (1934); *Chem. Abstr.* **33**, 2246 (1939).
61. M. Matsumoto and M. Maeda, *Kobunshi Kagaku* **12**, 428 (1955).
62. S. Nigam and S. Banerjee, *Indian J. Chem.* **4**, 301 (1966).
63. D. Braun and K. H. Becker, *Ind. Eng. Chem., Prod. Res. Dev.* **10**, 386 (1971).
64. S. Molnar, *J. Polym. Sci., Part A-1*, **10**, 2245 (1972).
65. M. Imoto, T. Otsu, and T. Ito, *Bull. Chem. Soc. Jpn.* **36**, 310 (1963).

66. K. Yonetani and W. W. Graessley, *Polymer* **11**, 222 (1970).
67. J. A. Vona, J. R. Constanza, H. A. Cantor, and W. J. Roberts, *in* "Manufacture of Plastics" (W. M. Smith, ed.), Vol. 1, p. 216, Van Nostrand-Reinhold, Princeton, New Jersey, 1964.
68. L. M. Minsk and E. W. Taylor, U.S. Patent 2,582,055 (1952); *Chem. Abstr.* **46**, P3800h (1952).
69. M. Shoichi, K. Nishioka, and K. Sato, Japanese Patent 70/03,389 (1970); *Chem. Abstr.* **72**, 112065s (1970).
70. A. M. Florescu and N. Cobianu, *Mater. Plast. (Bucharest)* **29**(1), 10 (1992).
71. M. Bravar, J. S. Rolich, N. Ban, and V. Gnjatovic, *J. Polym. Sci., Polym. Symp.* **47**, 329 (1974).
72. B. F. Greek, *Chem. & Eng. News* **55**, 8, Sept. 19 (1977).
73. R. G. R. Bacon, *Q. Rev., Chem. Soc.* **9**, 287 (1955).
74. L. Shapiro, *Tappi* **40**(5), 387 (1957).
75. S. Okamura and T. Motoyama, *J. Polym. Sci.* **58**, 221 (1962).
76. H. Naidus, *Ind. Enq. Chem.* **45**, 714 (1952).
77. A. S. Dunn and P. A. Taylor, *Makromol. Chem.* **83**, 207 (1965).
78. H. W. Starkweather and B. B. Taylor, *J. Am. Chem. Soc.* **52**, 4708 (1930).
79. K. Chujo, Y. Harada, S. Tokuhara, and K. Tanaka, *J. Polym. Sci., Part C* **27**, 321 (1969).
80. M. Nomura, M. Harada, W. Eguchi, and S. Nagata, *Polym. Prepr., Am. Chem. Soc., Div. Polym. Chem.* **16**(1), 217 (1975).
81. W. S. Port, J. E. Hansen, E. F. Jordan, Jr., T. J. Dietz, and D. Swern, *J. Polym. Sci.* **7**, 207 (1951).
82. H. Hopff and J. Dohany, *Makromol. Chem.* **69**, 131 (1963).
83. "Vynate Copolymer Latexes For Exterior Architectural Coatings," Technical Bulletin F-60831, Union Carbide Chemicals and Plastics Co., Danbury, Connecticut, 1994.
84. P. S. Martin, O. W. Smith, and D. R. Bassett, *Proc. Am. Chem. Soc., Polym. Mater. Sci. Eng.* **64**, 273 (1991).
85. K. E. J. Barrett and M. W. Thompson, *in* "Dispersion Polymerization in Organic Media," (K. E. J. Barrett, ed.), Wiley (Interscience), New York, 1975.
86. D. W. J. Osmond, British Patent 1,052,241 (1966); German Patent 1,520,119; *Chem. Abstr.* **69**, 10929y (1967).
87. L. A. Pilato and E. R. Wagner, French Patent 1,531,022, 1968; *Chem. Abstr.* **71**, 13770s (1969).
88. J. V. Dawkins and S. A. Shakir, *Proc. Am. Chem. Soc., Polym. Mater. Sci. Eng.* **64**, 357 (1991).
89. C. Gardner Swain and P. D. Bartlett, *J. Am. Chem. Soc.* **68**, 2381 (1946).
90. M. S. Matheson, E. E. Auer, E. B. Bevilacqua, and E. J. Hart, *J. Am. Chem. Soc.* **71**, 2610 (1949).
91. W. I. Bengough and H. W. Melville, *Proc. R. Soc. London, Ser. A* **225**, 330 (1954).
92. J. W. Breitenbach and H. F. Kauffmann, *Makromol. Chem.* **162**, 295 (1972).
93. I. M. Barkalov, V. I. Goldanskii, N. S. Enikolopyan, S. F. Terekhova, and G. M. Trofimova, *J. Polym. Sci., Part C* **4**, 909 (1963).

93a. Y. Tsuda, *J. Polymer Sci*. **49**, 369 (1961).
94. D. Mardare and K. Matyjaszewski, *Am. Chem. Soc., Polym. Prepr.* **34**(2), 566 (1993).
95. N. Ashikari, *J. Polym. Sci.* **28**, 250 (1958).
96. K. Noro and H. Kawazura, *J. Polym. Sci.* **45**, 264 (1960).
97. S.-I. Nozakura, M. Sumi, M. Uoi, T. Okamoto, and S. Murahashi, *J. Polym. Chem., Part A-1* **11**, 279 (1973).
98. T. Yatsu and H. Maki, Japanese Patent 74 32,669 (1974); *Chem. Abstr.* **82**, P73690c (1975).
99. G. E. Serniuk and R. M. Thomas, U.S. Patent 3,183,217 (1965); British Patent 946,052 (1964).
100. R. B. Seymour, D. P. Garner, G. A. Stahl, and L. J. Sanders, *Polym. Prepr. Am. Chem. Soc., Div. Polym. Chem.* **17**(2), 660 (1976).

24
POLYMERIZATION OF ALLYL ESTERS

Compared to the reactivity of acrylic esters, the polymerization of allyl alcohol and its esters is generally sluggish. Consequently, the syntheses of allyl esters are quite straightforward without the hazards of unwanted polymerizations during the monomer preparation. Consequently, many allyl esters are accessible. However, only a very few are of industrial significance; of those, most of them find application in specialized copolymerizations.

The free-radical polymerization process with allyl esters resembles that of vinyl and acrylic esters in that there are similar initiation, propagation, and termination steps. Of particular importance in the case of allyl compounds is the chain-transfer process. It is this chain-transfer reaction which constitutes the predominant characteristic differences between allylic and vinyl polymerization reactions. In the case of allyl esters both "effective" and "degradative" chain-transfer seem to occur side by side.

The polymerization of allyl methacrylate and acrylate is of more than academic interest. The reactivities of allyl and acrylic moieties are so different that the acrylic portion of the monomer essentially may homopolymerize while the allylic groups are unaffected. At a latter stage, perhaps by increasing the reaction temperature or by incorporating additional initiator, the allyl groups can be used to crosslink the polymer. Such a two-stage system permits ready mold-

---

From S. R. Sandler and W. Karo, "Polymer Syntheses," 2nd ed., Vol. III, pp. 282ff, by permission of Academic Press, San Diego, 1994.

ing or extrusion of an acrylic copolymer which can be rendered thermosetting at a later time.

Of particular interest is diethylene glycol bis(allyl carbonate), which, as "CR-39," is widely used in plastic spectacle lenses to provide exceptional clarity and scratch, impact, and abrasion resistance.

The polymerization of diallyl o-phthalate as well as of other α-, ω-diallyl esters is characterized by the formation of cyclic structures along the low molecular weight chains. At approximately 25% conversion to a low molecular weight polymer, the mass rapidly gels. The ease of ring formation of diallyl esters decreases with increases in the separation between the two allyl groups. Interestingly enough, the cyclic reaction involved in diallyl o-phthalate prepolymers form quite readily even though the rings probably consist of 11- or 12-member rings, sizes thought to be difficult to establish, at least in the case of cyclic polymethylene derivatives.

The diallyl esters of the o- and m-phthalic acids can be converted to prepolymers, which can be molded and then crosslinked. More usually, these prepolymers are added to other resin systems for subsequent formation of insoluble rigid plastics.

The prepolymers can also be dispersed in other monomeric systems and copolymerized. This can materially reduce the volume shrinkage when monomers are converted to a resin.

Diallyl chlorendate (I) has found application because of its high chlorine content. When added to other plastic materials, it is thought to improve their flame and fire resistance.

$$\begin{array}{c}\text{Cl}\\|\\\text{C}\end{array}$$

Cl—C⟨ ⟩CH—C(=O)—O—CH$_2$—CH=CH$_2$
‖ Cl—C—Cl |
Cl—C⟨ ⟩CH—C(=O)—O—CH$_2$—CH=CH$_2$
   C
   |
   Cl

Diallyl chlorendate (I)

References [1–13] are a selection of review articles and books dealing with the chemistry of allyl compounds.

Below, preparations of polymers based on typical allylic monomers are presented.

1. POLYMERIZATION OF ALLYL ACRYLATE AND METHACRYLATE

Allyl acrylate and methacrylate are reasonably simple to synthesize. In fact, allyl methacrylate is manufactured in bulk quantities. Since each of these monomers contains two double bonds of distinctly different susceptibility to polymerization when considered individually, these monomers are interesting both from the practical as well as the theoretical standpoint.

Early interest in allyl methacrylate arose from the concept that it was possible that this monomer would copolymerize with methyl methacrylate through its methacrylate bonds. The resulting resin could then be shaped, for example by injection molding, since it was still thermoplastic. After the molding had been completed, a separate heating cycle would bring about crosslinking through the allyl bonds to produce hard thermoset materials. Early applications in the production of plastic dental prostheses were visualized.

One of the early theoretical studies of the polymerization of allyl acrylate considered the "homopolymerization" of this monomer to be an "intramolecular copolymerization" of the allyl and the acrylic double bonds. In this process it was calculated that only about 3% of the allyl groups participated in the process. Upon extended heating, the residual allyl groups served as crosslinking sites [14].

An examination of the earlier allyl methacrylate literature by Butler [15] indicated to him that the results reported by earlier investigators can be explained, in part, by assuming that cyclization of allyl acrylate or methacrylate takes place to some extent before gelation takes over.

The bulk polymerization of allyl acrylate with benzoyl peroxide as initiator to 10% conversion gave a brittle, glassy polymer which was considered partially cyclized with a K_c (ratio of rate of cyclization to the rate of bimolecular propagation) of 0.41 mole/liter [16]. The polymer was described as soluble in both toluene and in carbon tetrachloride.

Two possible structures of the cyclic radicals which may be involved in the polymerization of allyl methacrylate are shown here:

$$
\begin{array}{cc}
\underset{(a)}{R-CH_2-\overset{\overset{\displaystyle CH_3}{|}}{\underset{\underset{\displaystyle O=C}{|}}{C}}\overset{\displaystyle CH_2}{\underset{\displaystyle \diagdown O \diagup}{\diagup}}\overset{\displaystyle \diagdown}{\underset{\displaystyle CH_2}{CH\cdot}}} & \underset{(b)}{R-CH_2-\overset{\overset{\displaystyle CH_3}{|}}{\underset{\underset{\displaystyle O=C}{|}}{C}}-\underset{\underset{\displaystyle O}{\diagdown \diagup}}{CH}-\underset{\displaystyle CH_2}{CH_2\cdot}}
\end{array}
$$

These are structures of cyclic allyl methacrylate free radicals [17]. (a) δ-lactone unit with carbonyl stretch frequency at 1740 and 1230 cm^{-1}; (b) γ-lactone unit with carbonyl stretch frequency at 1775 and 1275 cm^{-1}.

Procedure 24-1 is an example of the solution polymerization of allyl methacrylate using benzoyl peroxide as initiator. The polymer is said to be thermoplastic. After the primary, noncrosslinked polymer has been isolated and dried, some additional initiator may be blended with the resin. Under pressure and with cautious heating, a transparent sheet may be formed.

24-1. Solution Polymerization of Allyl Methacrylate [18]

In a 100 ml, round-bottom flask fitted with a reflux condenser, a solution of 10 gm of allyl methacrylate and 0.7 gm of recrystallized benzoyl peroxide in 56 gm of dry acetone is heated at reflux for 1.5 hr. The solution is then cooled and carefully poured with vigorous stirring into 1 liter of methanol. The precipitated polymer is collected on a filter, dried under reduced pressure at room temperature. The product is fusible and soluble in acetone. Upon heating under a slight pressure at 90°C the polymer fuses and converts to an insoluble, infusible material.

2. POLYMERIZATION OF DIALLYL CARBONATES

The monomer diethylene glycol bis(allyl carbonate), as shown here:

$$\left(CH_2=CH-CH_2-O-\overset{\overset{\displaystyle O}{\|}}{C}-O-CH_2-CH_2 \right)_2 O$$

has considerable commercial value as a highly transparent, hard, scratch-resistant resin for use in high-quality plastic lenses. Both the monomer and polymer have been designated by the trade name "CR-39." The monomer is also referred to as "allyl diglycol carbonate." While other diallyl carbonate derivatives have been studied [19], the primary interest has been in the CR-39 monomer and its polymerization. Like most allyl esters, diethylene glycol bis(allyl carbonate) requires high concentrations of initiators for conversion. Even then, to complete the process, extensive postcures are required.

Since the monomer is reasonably nonreactive, solutions of up to 5% of benzoyl peroxide in CR-39 monomer may be stored at 10°C or lower. If diisopropyl peroxydicarbonate is used, monomer solutions of this material must be stored below $-5°C$. The polymerization is air inhibited.

With care, a noncrosslinked syrup can be prepared which finds application as an optical cement. Whether this material is a cyclic prepolymer seems not to have been considered. Such structures may have rings with 16 members.

Procedure 24-2 is an early example of the preparation of a thermoplastic resin in solution.

24-2. Solution Polymerization of Diethylene Glycol Bis(allyl carbonate) [19]

In a four-necked, 500 ml flask fitted with a reflux condenser, mechanical stirrer, thermometer, and a means of maintaining a nitrogen atmosphere, are placed 100 gm of diethylene glycol bis (allyl carbonate), 100 gm of dioxane, and 4 gm of dibenzoyl peroxide. The mixture is stirred and the air is displaced with nitrogen. Then the reaction solution is heated between 80° and 85°C until a noticeable increase in viscosity is observed. The reaction mixture is then cooled to room temperature. Methanol is added until the solution becomes slightly turbid. Then the turbid mixture is added with vigorous stirring to five times its volume of methanol. The polymer is filtered off and dried under reduced pressure. This granular, white product, on mixing with 5 wt % of dibenzoyl peroxide, at 145°C and with the application of a pressure of 13.79 MPa (2000 psi), produces a transparent, infusible sheet.

To prepare glazing, transparent sheets are cast in cells made of plate glass with flexible gaskets used as spacers (cf. Sandler and

Karo [20], Procedure 4-2, for a detailed description of a glass casting cell). Control of the polymerization temperature used with such cells varies considerably with the thickness of the polymer sheet to be produced. For example, a sheet 3 to 5 mm thick can be prepared at 70°C with a monomer solution containing 3% of dibenzoyl peroxide over a period of 60–72 hr in a circulating air oven. After the sheet is removed from the cell, it is postcured at 115°C for an additional 2 hr. With thicker sheets, lower initial temperatures are required to permit proper dissipation of the heats of polymerization.

3. POLYMERIZATION OF DIALLYL ESTERS OF PHTHALIC ACIDS

The dially esters of o-phthalic and of isophthalic acids are commercially available both as monomers and as low molecular weight prepolymers. Diallyl terephthalate has been studied only occasionally. Diallyl chlorendate is of interest in copolymer systems since its high chlorine content is thought to contribute to the flame resistance of organic materials.

The most extensive research and development activity has involved diallyl o-phthalate. Diallyl isophthalate polymerizes more rapidly than the *ortho*-isomer. It cyclizes less during the early stages of polymerization. Consequently the prepolymer of the isophthalate has more reactive double bonds available for further reaction than the o-phthalate and the final resin produced from it is more highly crosslinked [21].

The methods of polymerization of the diallyl phthalates deal with control of the process to permit isolation of the thermoplastic prepolymer before the fully cured, crosslinked resin is formed.

Simply heating diallyl o-phthalate with benzoyl peroxide at 115°–125°C produced a highly crosslinked thermoset resin [22]. The usual methods of producing "prepolymers" involve the interruption of the polymerization process before gelation sets in (at about 25% conversion).

One novel method of polymerization without the use of any peroxide initiator involves heating diallyl o-phthalate under nitrogen in the presence of metallic copper. Below 205°C, copper acts as a retarder of the polymerization. However, above 225°C it accelerates the process (at about 215°C it neither inhibits nor accelerates) [23].

Shokal and Bent [23] pointed out that the course of the polymerization may be monitored by taking the refractive index of the solution of the prepolymer in the monomer. The refractive index of the pure monomer is n_D^{25} 1.5185. For each 1% of soluble polymer formed, the refractive index increases by 0.0005 units. Thus, at 25% conversion the refractive index of the solution is approximately n_D^{25} 1.531.

The isolation of the prepolymer from the reaction mixture usually involves its precipitation with an alcohol or some other nonsolvent. Two Japanese patents are based on this well-known phenomenon. In one, 100 gm of a prepolymer solution containing 25% of the prepolymer is treated at 40°C with 200 gm of ethanol in an extractor operating at 200 rpm for 5 min. After a second extraction at 65°C for 5 min with 240 gm of ethanol, 25 gm of the white prepolymer is isolated. Propanol, isobutanol, and isopropanol may also be used. The process is also applicable to poly(diallyl isophthalate) [24]. For poly(diallyl terephthalate) the use of methanol at 40°C followed by a second extraction with methanol at 65°C is patented [25].

Using high-temperature initiators such as dicumyl peroxide or *tert*-butyl perbenzoate at high temperature, the expected oxygen inhibition from the environment is substantially reduced and the polymerization proceeds at a reasonable rate [26].

The use of a high-temperature initiator actually goes back to a 1947 patent. In that patent, di(*tert*-butyl)peroxide is used as an initiator at 65°C. This is curious since the half life of this peroxide at 100°C is over one week! We wonder whether this may not be a typographical error. At 135°C, for example, the half life would be about 4 hr, which would be much more reasonable. However, in Procedure 24-3 we follow the original directions [27].

24-3. Bulk Polymerization of Diallyl *o*-Phthalate with High-Temperature Initiator [27]

In a 1 liter flask fitted with mechanical stirrer, a nitrogen bleed, a means of withdrawing samples, and a thermometer is placed 100 gm of diallyl *o*-phthalate and 2 gm of di(*tert*-butyl)peroxide. The mixture is heated with stirring under nitrogen at 65°C until the refractive index of the solution reached n_D^{25} 1.5313. Then the reaction mixture is slowly poured, with stirring, into 600 ml of methanol. The semisolid prepolymer is filtered off and dried under reduced pressure.

The product is soluble in a mixture of 3 parts of toluene and 1 part of xylene. This solution may be used as a bake-on coating on steel at 150°C for 1 hr. A hard, flexible, transparent, water-white film forms.

The prepolymer has also been prepared in two stages at two different temperatures using two initiators which operate at two distinctly different temperatures like *tert*-butyl hydroperoxide and di-*tert*-butyl peroxide [28]. In a strictly thermal process, diallyl *o*-phthalate has been polymerized at 200°–250°C. The conversion of monomer to polymer was followed by checking the change in refractive index with time. The process was "short stopped" before the gel point was reached by adding a solvent which separated unreacted monomer from the polymer [29].

Usually, suspension polymerizations consist of a process in which droplets of a monomer (or monomers) containing an initiator in solution are polymerized while being dispersed in an aqueous medium containing a suspending agent. Attempts are usually made to carry the process to high conversion and to isolate the product as rigid beads. However, in the case of the polymerization of the diallyl phthalates, any process which leads to essentially complete conversion will result in crosslinked beads which will have virtually no uses. Therefore a suspension process for the present monomers would have to be carried out to a modest conversion and isolation of the prepolymer would have to be from the solution of the polymer in its monomer.

With dimethylbenzyl alcohol as a polymerization regulator which prevents crosslinking and using animal glue as a suspending agent, a "suspension" polymer has been prepared. This material gave clear solutions in acetone and could be molded as 85°C and 5000 psi within 10 min. Procedure 24-4 illustrates the polymerization process of this patented process [30].

24-4. "Suspension Polymerization" of Diallyl *o*-Phthalate [30]

In a 2 liter resin kettle filled with an addition funnel, reflux condenser, mechanical stirrer, thermometer, and a means of maintaining an inert atmosphere, to 600 gm of water heated, with stirring, under a nitrogen atmosphere at 80°C, are added 2 gm of animal glue and 35 gm of dimethylbenzyl alcohol. While maintaining a temperature of 80°C, a solution of 15 gm of recrystallized dibenzoyl peroxide in 150 gm of diallyl o-phthalate is added from the addition funnel while stirring vigorously. After heating for 21 hr, the mixture is cooled to 30°C and transferred to a separatory funnel.

The lower organic layer is added dropwise with vigorous stirring to 500 ml of methanol. The precipitated prepolymer is separated, resuspended in fresh methanol, filtered, and dried under reduced pressure.

The yield is 76 gm with a saponification number of 428, an iodine number of 58, and a viscosity, in a 5% solution in benzene, of 0.82 cPs.

The use of magnesium carbonate at levels of 2–5% based on diallyl o-phthalate has been suggested as a suspending agent. Since the abstract indicates that polymer conversions of greater than 94% are achieved, we presume that this suspending agent was only of use when rigid polymer beads were to be isolated. At that level of conversion, the polymer may be expected to be thoroughly crosslinked. The same suspending agent had also been suggested for the polymerization of triallyl citrate [31].

Aqueous emulsions containing between 30% and 70% of the prepolymer of mixtures of monomer and prepolymer have been suggested for application in textiles and paper making. The emulsion is said to be formulated with 1% of *tert*-butyl perbenzoate. The composition is cured under pressure at 140°–160°C. Details on the preparation of a true latex are not available [32].

The soluble prepolymers of diallyl isophthalate may be cured not only by conventional thermal methods but also by photocrosslinking techniques. The use of aryl diazides as photoinitiator of solutions of poly(diallyl isophthalate) in aryl ketones has been suggested. With synergistically active sensitizers such as benzil or benzophenone with Michler's ketone curing is possible at wavelengths between 300 and 400 nm [33].

Studies of the pre-gel stage of free radical polymerizing diallyl isophthalate as well as of diallyl sebacate have been shown to involve significant intramolecular cyclization. On the average six to eight monomer units appear to be involved in each cycle [34, 35].

REFERENCES

1. W. Karo, *in* "Encycl. Ind. Chem. Anal.," Vol. 5, pp. 75–109, Wiley, New York, 1967.
2. R. C. Laible, *Chem. Rev.* **58**, 807 (1958).
3. R. C. Laible, *Encycl. Polym. Sci. Technol.* **1**, 750 (1964).
4. F. P. Greenspan, H. H. Beachem, and R. L. McCombie, *Encycl. Polym. Sci. Technol.* **1**, 785 (1964).
5. H. Reach, Jr., "Allyl Resins and Monomers," Van Nostrand-Reinhold, Princeton, New Jersey, 1965.
6. V. I. Volodina, A. I. Tarasov, and S. S. Spasskii, *Usp. Khim.* **39**(2), 276 (1970); *Chem. Abstr.* **73**, R4193v (1970).
7. C. E. Schildknecht, "Allyl Compounds and Their Polymers (Including Polyolefins)," Wiley (Interscience), New York, 1973.
8. W. Krolikowski and I. Prusinska, *Polimery (Warsaw)* **18**(1), 1 (1973); *Chem. Abstr.* **79**, T54099p (1973).
9. A. Jefferson and H. H. Kippo, *Proc. R. Aust. Chem. Inst.* **41**(6), 129 (1974).
10. M. Oiwa, *Parasuchikkuso* **25**(12), 13 (1974); *Chem. Abstr.* **82**, R86652s (1975).
11. M. Oiwa and A. Matsumoto, *Kobunshi* **25**(6), 387 (1976); *Chem. Abstr.* **85**, R47105h (1976).
12. C. E. Schildknecht, *in* "Concise Encyclopedia of Polymer Science and Engineering" (J. I. Kroschwitz, ed.), p. 262ff, Wiley, New York, 1990.
13. R. Dowbenko, *in* "Kirk–Othmer Encyclopedia of Chemical Technology," 4th ed., Vol. 2, pp. 161ff, Wiley, New York, 1992.
14. L. Grindin, S. Medvedev, and E. Fleshler, *Zh. Obshch. Khim.* **19**, 1694 (1949); *Chem. Abstr.* **44**, 1020a (1950).
15. G. B. Butler, *J. Polym. Sci.* **48**, 279 (1960).
16. M. Raetzsch and L. Stephen, *Plaste Kautsch.* **18**(8), 572 (1971); *Chem. Abstr.* **76**, 46545 (1972).
17. J. P. J. Higgins and K. E. Weale, *J. Polym. Sci., Part A-1*, **6**, 3007 (1961).
18. M. A. Pollack, I. E. Muskat, and F. Strain, U.S. Patent 2,273,891 (1942); *Chem. Abstr.* **36**, 3878 (1942).
19. I. E. Muskrat and F. Strain, U.S. Patent 2,592,058 (1952).
20. S. R. Sandler and W. Karo, "Polymer Syntheses," 2nd ed., Vol. 1, p. 333, Academic Press, San Diego, 1992.
21. C. F. Schildknecht, "Allyl Compounds and Their Polymers (Including Polyolefins)," p. 361, Wiley (Interscience), New York, 1973.
22. T. F. Bradley, U.S. Patent 2,311,327 (1943); *Chem. Abstr.* **37**, 4502 (1943).
23. F. C. Shokal and F. A. Bent, U.S. Patents 2,475,296 and 2,475,297 (1949).
24. T. Tanaka and S. Takayama, Japanese Patent 70/14,546 (1970); *Chem. Abstr.* **73**, 110442b (1970).
25. T. Tanaka and S. Takayama, Japanese Patent 70/14,547 (1970); *Chem. Abstr.* **73**, 110443n (1970).
26. C. L. Wright and H. H. Beacham, U.S. Patent 3,527,665 (1970); French Patent 1,551,69? (1970).
27. W. E. Vaughan and F. F. Rust, U.S. Patent 2,426,476 (1947).
28. E. C. Shokal, British Patent 604,544 (1948).

29. J. K. Wagner and E. C. Shokal, U.S. Patent 2,446,314 (1947); *Chem. Abstr.* **42**, 852 (1948).
30. C. A. Heiberger and J. L. Thomas, U.S. Patent 2,832,758 (1958).
31. K.-S. Chia and F.-Y. Chao, *Chemistry (Taiwan)* p. 31 (1956); *Chem. Abstr.* **51**, 1645 (1957).
32. W. Festag, German Patent 1,163,774 (1964); *Chem. Abstr.* **60**, 1345z (1964); Solvay, Belgian Patent 562,701 (1959); *Chem. Abstr.* **53**, 10845 (1959).
33. M. N. Gilano and M. A. Lipson, *Tech. Pap., Reg. Tech. Conf., Soc. Plast. Eng., Mid Hudson Sect.* p. 30 (1970); *Chem. Abstr.* **74**, 4077f (1971).
34. I. I. Romantsova, *Int. Polymer. Mater.* **19**(1–2), 51 (1993).
35. I. I. Romantsova, *Polymer. Mater. Sci. Eng.* **66**, 121 (1992).

25
POLY(VINYL ALCOHOL)

Poly(vinyl alcohol) is widely used in many industries. It finds application in films, coatings, fibers, and as a viscosity modifier of a variety of aqueous systems such as certain cosmetics. Its films, formulated in subdued light with alkali metal dichromate salts, may be crosslinked by exposure to ultraviolet light. This property has found application in photoengraving and related fields [1].

The technology of the industrial production of poly(vinyl alcohol) is quite complex. Much of it is guarded as in-house know-how. Suffice it to say that commercial PVAlc is available in a variety of molecular weights, degrees of hydrolysis, degrees of hydrolysis of the starting poly(vinyl acetate), average degree of branching of the polymer chains, variations in molecular weight distribution, distribution of monomer units along the polymer chains, and the level of inorganic residues from the manufacturing process. The latter factor is of particular importance if the PVAlc is to be used in cosmetic, ophthalmic, or medicinal applications.

Poly(vinyl alcohols) can be considered as block copolymers of poly(vinyl acetate) and poly(vinyl alcohol). The resins may be crosslinked to some extent as will be discussed below. There may also be branched polymer chains.

The degree of hydrolysis of the starting poly(vinyl acetate) and the molecular weight of the material are the usual specifications used in describing these products. The degree of hydrolysis may be determined by conventional saponification procedures. The molecular weight is usually based on viscosity measurements at 20°C of 4% solutions in deionized water.

It is interesting to note that PVAlc, which contains approx. 20% poly(vinyl acetate), is more easily dispersed in water that "100%"

PVAlc. It has been suggested that at this ratio of the two monomeric units, the balance of hydrophilic to hydrophobic characteristics is particularly favorable.

Monomeric vinyl alcohol exists only in its tautomeric form as acetaldehyde. Therefore, poly(vinyl alcohol) must be prepared by an indirect method such as the hydrolysis of polymers of vinyl esters. For practical reasons, the starting material of choice is poly(vinyl acetate). While hydrolysis may be carried out under acidic conditions, alkaline conditions in the presence of an alcohol are preferred. The reaction may be represented by the following equation:

$$\text{+CH}_2\text{CH[OCOCH}_3\text{]+}_n + n\text{CH}_3\text{OH} \longrightarrow \text{+CH}_2\text{CH[OH]+}_n + n\text{CH}_3\text{OCOCH}_3$$

To be noted is that methyl acetate is a coproduct of this process. As a matter of fact, in industrial processes, this product is isolated by distillation as a methanol–methyl acetate azeotrope.

The base-catalyzed alcoholysis of poly(vinyl acetate) is quite rapid and is thought to be autocatalytic. Under the usual reaction conditions, the reaction goes to approximately 90% completion. To reach 100% conversion requires specialized conditions. To achieve partial hydrolysis is also difficult because of the rapidity of the process.

The rate of hydrolysis of poly(vinyl acetate) (PVAc) is independent of the molecular weight of the starting material. The degree of hydrolysis also seems to be independent of the molecular weight of the starting poly(vinyl acetate).

Commercial poly(vinyl alcohol) is available in several degrees of hydrolysis as well as in several molecular weight ranges. The PVAlc with a degree of hydrolysis of approximately 81% is of particular interest. This material appears to be the most water soluble of the commercial products.

Note that to dissolve PVAlc in water is, at best, troublesome. To avoid lump formation, it is best to sprinkle the product very slowly into rapidly stirred water. Even then great care and patience are needed. For viscosity measurements in aqueous solution, concentrations of 4% polymer in water are usual. This may well be the limit of the solubility of this resin.

Predicting the molecular weight of the hydrolysis product is problematical. Most commercial PVAc is somewhat crosslinked.

The crosslinks may involve both the methylene group of the vinyl unit, which may be anticipated, or the methyl of the acetate moiety, which is unusual. Base hydrolysis of the former type of crosslink does not affect the chain length of the polymer. The hydrolysis of an ester linkage whose methyl group is attached to another polymeric group will result in the cleavage of the crosslinked chain. Therefore there would be a reduction of the anticipated average molecular weight of the PVAlc that is being formed, if a "methyl crosslink" is present. Then, upon reacetylation of PVAlc with acetic anhydride, a PVA with lower molecular weight will form. This product is, presumably, not crosslinked. On alcoholysis of this second product, a PVAlc forms that has substantially the same molecular weight as the PVAlc that had been acetylated just prior to this last alcoholysis. These phenomena have been observed, leading to the conclusion that indeed there are two types of crosslinks [2–5].

A sample of poly(vinyl alcohol) which had been freed of crosslinks through the ester portions of the starting material lends itself to an evaluation of the ratio of head-to-tail chains to head-to-head, tail-to-tail chains. If this latter structural feature is present, there should be *gem*-diol structures corresponding to each head-to-head unit in a sample of poly(vinyl alcohol).

It is well known that periodic acid is a specific reagent for the cleave of *gem*-diols. If this reagent is applied to an aqueous solution of poly(vinyl alcohol), chain-scission would take place at the head-to-head moieties. This would result in a reduction in the overall molecular weight of the polymer. By measuring the viscosity molecular weight of the PVAlc before and after treatment with periodic acid, the ratio of the two structural features can be calculated [6].

In the preparation described here, the alkaline reagent of choice is potassium hydroxide because it is more soluble in methanol than sodium hydroxide. In the final purification of the product, residues of potassium salts and of potassium hydroxide are also more easily removed than the sodium analogs. Commercially, sodium methoxide is often used as the catalyst for the hydrolysis.

25-1. Preparation of Poly(vinyl alcohol) by the Alcoholysis of Poly(vinyl acetate)

With suitable safety precautions and equipment, including rubber gloves, rubber apron, safety goggles, and a face shield, in a mortar, cautiously and rapidly grind 0.5 gm of anhydrous potassium hydroxide to a fine powder with a pestle, transfer the powder to a 250 ml Erlenmeyer flask, add 100 ml methanol, stopper the flask, and disperse the potassium hydroxide. Preserve the material until required.

In a 1 liter three-necked flask fitted with a mechanical stirrer, addition funnel, and a reflux condenser place 50 gm of poly(vinyl acetate) and 500 ml of anhydrous methanol. Close the flask with the addition funnel, turn on the condenser, and, with vigorous stirring, heat the mixture at reflux until the polymer has dissolved. Reduce the heat to a moderate rate of reflux and dropwise add 100 ml of the solution of potassium hydroxide in methanol through the addition funnel. By controlling the rate of addition of the potassium hydroxide solution, the reaction rate may be moderated. If necessary, adding more methanol, lowering the heating mantle, even external cooling may be needed to maintain control of the hydrolysis.

After the reaction has slowed down, continue heating at reflux for 2 hr (see Note).

Cool the flask. Remove stopcock grease from all the joints of the flask. With suction, filter off the poly(vinyl alcohol) that has formed. Then wash the product repeatedly with methanol until a sample of the methanol washings, diluted with deionized water, has a pH below 7.5 (i.e., until residual alkali has been removed).

Collect the product on a Büchner funnel, and air dry the product. The final drying of the product to constant weight may be carried out in a vacuum desiccator or in a vacuum oven at a moderate temperature.

NOTE: Because PVAlc is insoluble in methanol, it usually precipitates as it forms. Occasionally, a gel forms instead of a powder. After cooling, such gel may be broken up in a Waring blender with additional methanol and worked up as usual.

REFERENCES

1. D. L. Cincera, in "Kirk−Othmer Encyclopedia of Chemical Technology," 3rd ed., Vol. 23, p. 864, John Wiley and Sons, New York, 1983.
2. W. H. McDowell and W. O. Kenyon, J. Am. Chem. Soc. **62**, 415 (1940).

3. L. M. Minsk, W. J. Priest, and W. O. Kenyon, *J. Am. Chem. Soc.* **63**, 2715 (1941).
4. O. L. Wheeler, S. L. Ernst, and R. H. Crozier, *J. Polym. Sci.* **8**, 409 (1952).
5. S. R. Sandler and W. Karo, "Polymer Syntheses," 2nd ed., Vol. III, pp. 216ff, Academic Press, San Diego, 1996.
6. P. J. Flory and F. S. Leutner, *J. Polym. Sci.* **3**, 880 (1948).

26
INTRODUCTORY NOTES ON EMULSION POLYMERIZATION TECHNIQUES

1. INTRODUCTION

Polymerization procedures of vinyl-type monomers by either the suspension or the emulsion techniques closely resemble those conventionally used in the preparative organic synthesis laboratory. In some respects these polymerization procedures may be simpler than ordinary organic synthesis procedures.

A sharp distinction should be drawn between suspension (or slurry) polymerizations, on the one hand, and emulsion polymerizations, on the other.

Suspension polymerizations are processes in which water-insoluble monomers, containing a monomer-soluble initiator, are dispersed in water containing a suspending agent, and heated with rapid stirring, to initiate the process. Typical suspending agents are poly(vinyl alcohol), starches, gelatine, freshly precipitated calcium phosphate, salts of poly(acrylic acids), and various natural gums. The reaction kinetics of the process resemble that of many bulk polymerizations being carried out substantially simultaneously in drops that may have diameters from about 50 μm on up. The stirring rate during the reaction has a profound influence on particle size and shape. Since the monomer droplets are in reasonably good contact with a heat exchanger, water, the exothermic nature of the process is easily controlled. The resulting product may be isolated by filtration (unless some unwanted agglomeration had taken place). Plasticizers, pigments, chain-transfer agents, and various modifiers may be incorporated in the monomer droplets prior to the polymerization.

This process should be carefully differentiated from *emulsion polymerization* (also called *latex* polymerization). This involves the formation of colloidal polymer particles which are substantially permanently suspended in the reaction medium. The process involves the migration of monomer molecules from its liquid phase through the water to sites of polymerization consisting of miscelles containing surface-active agent molecules which surround monomer molecules. Generally, the process requires a water-soluble initiator in the presence of a surfactant. The polymer particles that form are of colloidal dimensions, i.e., on the order of between 0.1 and 1 μm [1]. Particles that have diameters greater than 1 μm usually settle out of the aqueous phase. Even so, they are often referred to as *latex particles* if they have been prepared by a typical emulsion polymerization procedure characterized by the use of a water-soluble initiator and/or an anionic surfactant.

The basic work on latex polymerization goes back about 50 years, to the World War II Synthetic Rubber Project. That project is a monument to enlightened research. It brought together a large group of scientists who were able to do both the basic physical chemical research along with the applied research to come up rapidly with the synthetic rubber that was so sorely needed during the war. Reference [2] shows how thoroughly at least one aspect of emulsion polymerization chemistry was studied during this period. It would almost appear that little has been added to that work from the preparative standpoint since that time.

It is beyond the scope of this work to discuss theoretical aspects of emulsion polymerization. References [1–15] contain material pertinent to the theoretical and practical aspects of this topic. While most of these references deal primarily with the Harkins–Smith–Ewart approach to the interpretation of the emulsion polymerization process, alternative mechanisms have been proposed [1]. Of particular interest in this connection is a recent examination of the role that the surfactant plays in the process. It has been known to most workers in the field that the surfactant influences the final properties and applications of a latex. However, rarely has the paradox been discussed of how a negatively charged initiator like the sulfate radical ion (SO_4^-) can possibly enter a micelle consisting of an assemblage of negatively charged surfactant anions and solubilized monomer molecules. One would expect coulombic repulsive forces to prevent this from taking place and therefore prevent initiation of polymerization [14].

2. DEVELOPING AN EMULSION POLYMERIZATION RECIPE

Water, monomer, surfactant, and an initiator are the basic ingredients of a latex polymerization recipe. Despite this apparent simplicity, Gerrens stated that the process involved at least 36 variables [5, 6]. There may actually be many more factors that enter into the problem of controlling emulsion polymerizations. Fortunately, many may be kept constant from preparation to preparation so that only the effect of the major factors need be considered.

A. EQUIPMENT

On a laboratory scale, standard glassware may be used unless low boiling monomers are involved. Then, pressure equipment will be required.

Resin flasks with four necks are convenient. Stirring with a stainless steel stirrer is usually recommended since this may contribute a catalytic amount of iron to the initiator system.

The stirring motor needs to operate only at such a speed that the monomers do not form a distinct layer in the flask. A speed of 150–200 rpm is usually adequate. However, a constant torque stirring motor is recommended.

Addition funnels, addition burets, condensers, and gas-inlet tubes are usually needed. While a thermometer in the resin flask may be adequate for monitoring the reaction temperature and a water bath may be suitable for heating, the use of a thermocouple with appropriate instrumentation is preferred. Such a thermocouple may be used with a controller to operate a hydraulic jack that raises or lowers a heating mantle as required. (Instruments for Research and Industry [I^2R] of Cheltenham, Pennsylvania, offers such equipment. I^2R also offers equipment for monitoring the stirring speed.) This information may be recorded with an X-Y recorder or its equivalent.

The replacement of the aqueous phase of a freshly prepared latex may be attempted by use of dialysis equipment. This is, at best, tedious and possibly ineffective. The use of a tangential flow filtration apparatus such as a "Pellicon" from Millipore or its equivalent may be more appropriate.

The methods of determining the particle sizes of the latex polymer are many. Many techniques used presuppose that the latex does not form a film under the conditions of the measurement. Transmission electron microscopy and disk centrifugation are among the techniques that may be considered.

B. MONOMER TO WATER RATIO

The ratio of the total nonvolatiles to water (usually referred to as *percent solids*) is important. Usually it is best to start experiments to develop the techniques required to prepare a 30–40% solids latex without the formation of coagula. Latices with higher percent solids are difficult to prepare. The geometry of the close packing of uniform spheres imposes a limitation on the percent nonvolatiles in the range of 60–65%. Dissolved nonvolatiles and the judicious packing of spheres of several diameters may permit the formation of more concentrated latices in principle.

Deionized water, preferably freshly boiled and cooled, is usually used in the polymerizations. The monomers need not necessarily be freed of inhibitors. If the monomer is to be deinhibited, this is most easily accomplished by passing the monomer through a short column packed with a coarse alumina such as Alcoa GS-20.

The need for an inert atmosphere in the reactor varies considerable with the nature of the monomers used and other factors. While the literature usually calls for the use of nitrogen, a heavier-than-air gas such as argon is recommended.

C. MONOMERS

Presumably most vinyl-type monomers may be used in the preparation of latices, provided that they are reasonably insoluble in water. Water-soluble monomers may be copolymerized with insoluble monomers in emulsion polymerization processes. Such systems are complex. Factors such as the distribution coefficients between the monomers with each other and with water come into play. The more water-soluble monomers may undergo solution polymerization to some extent and the resulting polymer may be adsorbed on other polymer particles. The reactivity ratios of the monomers are modified when an aqueous medium is involved. In fact, there may be further complications because some of the monomers may form hydrates prior to polymerization.

Among the monomers that undergo latex polymerization are styrene, styrene-butadiene, 1,2-dimethylene cyclohexane, various esters of acrylic and methacrylic acids, acrylonitrile, methacrylonitrile, acrylonitrile with comonomers such as vinyl acetate, methacrylonitrile, butadiene, vinylidene chloride, α-methylstyrene, vinyl chloride, and vinyl fluoride. Acrylic esters have been copolymerized with acrylonitrile as well as with acrylic or methacrylic acids. Other latices have been prepared by copolymerizing acrylamide with octylacrylamide or acrylamide with styrene. Fluorinated acrylamides have been copolymerized in emulsion recipes with acrylic esters. Biallyl has been emulsion copolymerized with various dithioacids [16].

Detailed examples of many of these emulsion polymerization systems are given in S. R. Sandler and W. Karo, "Polymer Syntheses," 2nd ed., Vols. I, II, and III, Academic Press, San Diego, California.

D. SURFACTANTS

Usually small amounts of a surfactant are added to the water prior to the addition of the monomer. To be sure, there are emulsion polymerizations that do not seem to make use of a deliberately added surfactant, as in the case of certain monodispersed poly(styrene) latices. However, on close examination of the reaction procedure, it will be noted that the initiator is usually a persulfate. The persulfate may initiate a small amount of polymerization. The polymer that is then formed has a sulfate terminal group. In this way, a poly(monomersulfate) surfactant may have been formed *in situ*.

Most emulsion polymerizations are carried out with anionic surfactants. For added stability of a latex or to create specific properties in the product, nonionic surfactants or other protective colloids may be added. Cationic surfactants usually do not lend themselves for latex polymerization. In fact they tend to coagulate polymer dispersions.

One of the earliest surfactant used was sodium stearate (Ivory soap flakes). In this connection it should be pointed out that care must be taken that little or no oleic acid or its salts are present. Oleic acid may act as an inhibitor or chain stopper for the polymerization.

Among surfactants that have been used is sodium dodecyl sulfate (SDS; sodium lauryl sulfate or SLS). In the case of this material, it is best that the surfactant be free of lauryl alcohol or other unsulfated carbinols. On the other hand, there is a grade available that is about 30% active in a yellowish aqueous solution. The discoloration is caused by iron oxide or iron salt dispersions. These may actually catalyze the initiation process.

Another commonly used material is sodium decylbenzene sulfate. Such surfactants as the various Tergitols, Triton X-100, Triton X-200, and various sodium bis(alkylsulfo)succinates (various aerosols) have been used along with many other industrial surface active agents. As protective colloids, typical materials are the poly(vinyl alcohols), various Spans, sorbitan esters, Arlcel-80, various hydroxyethyl cellulose, etc.

E. INITIATORS

Emulsion polymerizations are usually carried out with water-soluble initiators such as the salts of persulfuric acid. Since ammonium persulfate is not very shelf stable and potassium persulfate is somewhat difficult to dissolve in water, sodium persulfate may be the material of choice. Another initiator that has been suggested is 2,2'-azobisisobutyramidine · hydrochloride. These initiators are usually used at between 60 and 80°C.

So-called redox initiators consist of a system in which a reducing agent is added to the water−monomer−surfactant−oxidizing initiator system. Interestingly, in the presence of a monomer, the oxidizing agent and reducing agent initiate polymerization, usually at room temperature or even below, without seemingly destroying each other. Typical redox systems are sodium persulfate−sodium metabisulfite and hydrogen peroxide (30%)–tartaric or ascorbic acid.

F. REACTION PROCEDURE

The ratio of the various ingredients to each other (which Gerrens calls "Flottenverhältnis") [5] as well as the reaction temperature are significant in the development of an emulsion recipe.

In the preparation of a polymer latex, the initial relationship of water, surfactant, and monomer concentration determines the number of particles present in the reaction vessel. Once the process is under way, addition of monomer does not change the number of latex particles. If such additional monomer polymerizes, the additional polymer is formed on the existing particles. As expected, the smaller initial particles imbibe more of the additional monomer than the larger ones. Consequently a procedure in which monomer is added to preformed latex polymer tends to produce a latex with a uniform particle size, i.e., a "monodispersed latex." Since the stability of the latex is dependent to a major extent on the effective amount of surfactant on a particle surface, considerable increase of the volume of the latex particles is possible with minor increases of the surface area purely on geometric grounds (an increase of the volume of a sphere by a factor of 8, increases the surface area by a factor of 4, while the particle diameter only doubles). These considerations have many practical applications, not the least of which is the possibility of preparing latex particles started with one comonomer composition to which a different comonomer solution is added.

Note that the replacement of air in the reactor with nitrogen should be undertaken before surfactant is added to the water. Otherwise uncontrollable foaming may take place.

Also, toward the start of the polymerization, the reacting mixture changes in appearance to resemble milk with a sky-blue edge at the outer surface.

To improve the uniformity of the particles and to control the exothermic nature of the reaction, it is a common practice to initiate a latex polymerization with only about 10% of the available monomer with the full amount of water. Once the milklike appearance has been observed, the remaining monomer may be added.

G. EXAMPLES OF EMULSION POLYMERIZATION PROCEDURES

26-1. Emulsion Polymerization of Styrene [17]

To a resin kettle equipped with a mechanical stirrer, condenser, and nitrogen inlet tube is added 128.2 gm of distilled water, 71.2 gm of styrene, 31.4 ml of 0.680% of potassium persulfate, and 100 ml of 3.56% soap solution (see Note). The system is purged with nitrogen to remove dissolved air. Then the temperature is raised to 50°C and kept there for 2 hr to afford a 90% conversion of polymer. The polymer is isolated by freezing-thawing or by adding alum solution and boiling the mixture. The polystyrene is filtered, washed with water, and dried.

NOTE: In place of soaps such as sodium stearate one can use 1.0 gm of either sodium dodecyl benzenesulfonate or sodium lauryl sulfate.

26-2. Preparation of Butadiene-Styrene Copolymers by the Emulsion Polymerization Technique [18]

$$n\text{CH}_2=\text{CH}-\text{CH}=\text{CH}_2 + n\text{C}_6\text{H}_5-\text{CH}=\text{CH}_2 \longrightarrow$$
$$\left[\text{CH}_2-\text{CH}=\text{CH}-\text{CH}_2-\text{CH}_2-\underset{\underset{\text{C}_6\text{H}_5}{|}}{\text{CH}}\right]_n$$

To an ice–water–salt-cooled 275 ml stainless steel pressure bomb (or 4 oz bottles with metal caps containing a rubber gasket for a total charge of only 60 gm) is added 180 gm of distilled water, 5 gm of Proctor & Gamble SF Flakes (see Note a), 0.3 gm of potassium persulfate (reagent grade) (Note b), and 0.5 gm of dodecyl mercaptan dissolved in 25 gm of styrene. Then the butadiene, 75.0 gm, is weighted into an ice-chilled 4 oz bottle and one additional 0.5 gm is added to allow for transfer loss in pouring into the stainless steel bomb. The bomb is sealed and placed into a shaker–oil bath at 110°–125°C for 1 hr. The bomb is cooled in a stream of air and then finally cooled in an ice bath. The latex is poured out into a beaker containing phenyl-β-naphthylamine (see Note c) and coagulated with a saturated salt–dilute sulfuric acid solution to give 75 gm (75%) of 96% solubility in benzene.

NOTES: (a) Sodium soap of hydrogenated tallow fatty acid (anhydrous) may also be used. (b) A mixture of 0.08 gm of 100% active *p*-menthane hydroperoxide (Hercules Powder C.), 0.08 gm of sodium formaldehyde sulfoxylate ($FeSO_4 \cdot 7H_2O$), and 0.02 gm of Versene Fe-3 (100%) may be used if one wants the polymerization to proceed at 5°C [19]. In addition, amines and peroxides have been reported to give electron-transfer complexes which decompose at room temperature or below [20]. (c) **CAUTION:** carcinogenic.

The emulsion copolymerization of butadiene with acrylonitrile is of considerable industrial importance. Since butadiene is a gas at room temperature, laboratory experiments with this system require pressure equipment. A procedure which uses hydrogen peroxide (**CAUTION:** handle with care) as the source of free radical is described in Preparation 26-3.

26-3. Preparation of Acrylonitrile–Butadiene Copolymers by Emulsion Polymerization [21]

The preparation is carried out in a pressure vessel in an explosion-proof hood, with all due precautions, particularly in regard to explosive and fire hazards related to handling butadiene. To 250 gm of freshly boiled water, 5.0 gm of soap flakes (85% neutralized), 0.6 gm of diisopropyl xanthogen disulfide, 33.3 gm of acrylonitrile, and 66.7 gm of chilled liquid butadiene is added 0.3 gm of hydrogen peroxide (added as 30% hydrogen peroxide solution) and 0.1 gm of a mixture prepared by grinding together 1.57 gm of sodium pyrophosphate, 0.28 gm of ferric sulfate, and 0.0014 gm of cobaltous chloride.

A small amount of the butadiene is allowed to vaporize to displace the air in the vessel. Then the vessel is closed, the protective covering is placed over the vessel, and the assembly is rotated in a bath at 30°C. When the conversion of monomers to polymer has reached 70%, a 1% aqueous solution of hydroquinone is injected into the reactor. The reactor is cooled in an ice bath and cautiously opened. The excess butadiene is allowed to evaporate in the hood.

26-4. Emulsion Copolymerization of Styrene and Acrylamide [22, 23]

In a 300 ml flask fitted with a stirrer, a gas inlet tube and reflux condenser, 12 gm of styrene, 8 gm of acrylamide, and 150 ml of deionized water, buffered to a pH of 9, are subjected to a slow stream of nitrogen for 1 hr. Then 10 ml of a potassium persulfate (5 mmol per liter of water) is added. The mixture is heated, with stirring, at 70°C for 3 hr. The latex formed has particles of 0.31 μm diameter, conversion ca. 80%. The polymer may be isolated by pouring the latex into an excess of acetone, filtering the polymer off, and drying the product under the reduced pressure. We estimate that the resulting polymer contains ca. 20% acrylamide (i.e., ca. half of the charged acrylamide is not in the copolymer but probably was lost to the water layer).

26-5. Emulsion Polymerization of Vinyl Acetate without Surfactant [24]

In a 3 liter reaction vessel equipped with pressure-equalizing addition funnel, reflux condenser, thermometer, and mechanical stirrer, 3.5 gm of potassium persulfate and 5.5 gm of potassium citrate monohydrate are dissolved in 80 gm of water maintained at 82°C. To this solution, with agitation, 100 gm of vinyl acetate is added continuously over a 3 hr 20 min period. During the addition period the reaction temperature is maintained between 82 and 85°C. After the addition is complete, the temperature is allowed to rise and is maintained at 90°C for 20 min. Upon cooling, the yield of latex is 91.6%, the solids content 53%, and the particle size 0.3 ± 0.1 μm with remarkable uniformity of size. By conventional procedures, the expected particle size distribution would have been between 0.5 and 2 μm.

Procedure 26-6 is an adaption of a ter-polymerization that starts with the formation of a seed latex followed by the gradual addition of both a monomer composition and an initiator solution at separate rates. The resulting latex has a high percentage of nonvolatiles. It is said to be suitable for formulating good emulsion paints. In connection with this preparation, care must be taken that the initiating ammonium persulfate is indeed active.

26-6. Emulsion Ter-polymerization of Vinyl Acetate, Butyl Acrylate, and Vinyl Neodecanoate (Seeded Process with Gradual Monomer and Initiator Additions) [25]

To a 1 liter resin kettle equipped with a reflux condenser, an explosion-proof stirrer, a thermometer and temperature controller, and a nitrogen inlet is charged 204.00 gm of deionized water. Then 6.00 gm of Cellosize hydroxyethyl cellulose WP-300, 3.00 gm of Tergitol NP-40, 3.90 gm of Tergitol NP-15 (two nonionic surfactants), 3.3 gm of Siponate DS-4 (an anionic surfactant), and 0.6 gm of ammonium bicarbonate are added and the stirred mixture is blanketed with nitrogen and warmed to 55°C. This temperature is maintained for 20 min. A mixture of 18.00 gm of vinyl acetate, 4.50 gm of butyl acrylate, and 7.5 gm of vinyl neodecanoate is added, followed by 0.24 gm of ammonium persulfate. The reaction mixture is heated to 75°C and maintained at that temperature for 15 min to form the seed latex.

The reaction temperature is then raised to 78°C and the gradual addition of monomers and initiator solution is begun. The monomer solution, consisting of 162 gm of vinyl acetate, 41.4 gm of butyl acrylate, and 66.6 gm of vinyl neodecanoate, is added over a 2 hr period. The initiator solution of 0.60 gm of ammonium persulfate dissolved in 60.00 gm of deionized water is added over a period of 2.5 hr. All the while, the reaction temperature is maintained at 78°C.

After the last of the initiator solution has been added, heating and good stirring is continued for another hour. The latex is cooled and filtered through a 200 mesh stainless steel screen. The percent nonvolatiles of the latex is 53.2%. Coagulum is 0.01%.

26-7. Emulsion Polymerization of Ethyl Acrylate (Thermal Initiation) [26, 27]

In a 2 liter Erlenmeyer flask, in a hood, to 800 ml of deionized water is added in sequence 96 gm of Triton X-200, 1.6 gm of ammonium persulfate, and 800 gm of ethyl acrylate. The contents is thoroughly mixed to form a monomer emulsion.

In a 3 liter resin kettle fitted with a stainless steel stirrer, a thermometer which extends well into the lowest portion of the reactor, a reflux condenser, and a dropping funnel a mixture of 200 ml of deionized water and 200 ml of the monomer emulsion is heated in a water bath while stirring at a constant rate in the range of 160–300 rpm to an internal temperature of 80°–85°C. At this temperature refluxing begins and vigorous polymerization starts (frequently signalled by the appearance of a sky-blue color at the outer edges of the liquid). The temperature may rise to 90°C. Once refluxing subsides, the remainder of the monomer emulsion is added from the dropping funnel over a 1 to 2 hr period at such a rate as to maintain an internal temperature between 88° and 95°C by means of the hot water bath, if necessary. After the addition has been completed and the reaction temperature has subsided, the temperature of the latex is raised briefly to the boiling point. Then the reaction mixture is cooled to room temperature with stirring. After removing stopcock grease from all joints, the latex is strained through a fine-mesh nylon chiffon. Only a negligible amount of coagulum should be present.

In particular, if a latex is to be used for coatings, adhesives, or film applications, no silicone-based stopcock greases should be used on emulsion polymerization equipment. While hydrocarbon greases are not completely satisfactory either, there are very few alternatives.

In the above examples, it will be noted that the monomer is added to the reaction system as an oil-in-water emulsion. Many emulsion polymerizations are more simply carried out by adding pure monomer to an aqueous dispersion of surfactant and initiators. This procedure permits a more rigid control of the number of particles in the aqueous phase.

REFERENCES

1. F. W. Billmeyer, Jr., "Textbook of Polymer Science," 2nd, ed., Wiley (Interscience), New York, 1971.
2. F. A. Bovey, I. M. Kolthoff, A. J. Medalia, and E. J. Meehan, "Emulsion Polymerization," Wiley (Interscience), New York, 1955.
3. E. W. Duck, *Encycl. Polym. Sci. Technol.* **5**, 801 (1966); J. G. Brodnyan, J. A. Cala, T. Konen, and E. L. Kelley, *J. Colloid Sci.* **18**, 73 (1963).
4. H. Fikentscher, H. Gerrens, and H. Schuller, *Angew. Chem.* **72**, 856 (1960).
5. H. Gerrens, *Fortschr. Hochpolym.-Forsch.* **1**, 234 (1959).
6. H. Gerrens, *Ber. Bunsenges. Phys. Chem.* **67**, 741 (1963).
7. W. D. Harkins, *J. Amer. Chem. Soc.* **69**, 1428 (1947).
8. G. Odian, "Principles of Polymerization," McGraw-Hill, New York, 1970.
9. "Emulsion Polymerization of Acrylic Monomers," Bulletin CM-104 A/cf, Rohm and Haas Company, Philadelphia, Pennsylvania.
10. W. V. Smith and R. H. Ewart, *J. Chem. Phys.* **16**, 592 (1948).
11. J. W. Vanderhoff, *in* "Vinyl Polymerization" (G. E. Ham *et al.*, eds.), Vol. 1, Part II, p. 1, Dekker, New York, 1969.
12. P. J. Flory, "Principles of Polymer Chemistry," Cornell Univ. Press, Ithaca, New York, 1953; M. S. Guillod and R. G. Bauer, *J. Appl. Polym. Sci.* **16**, 1457 (1972).
13. C. P. Roe, *Ind. Eng. Chem.* **60**, 20 (1968).

14. A. S. Dunn, *Chem. Ind. (London)* p. 1406 (1971).
15. W. S. Zimmt, *J. Appl. Polym. Sci.* **1**, 323 (1959).
16. C. S. Marvel and E. A. Kraiman, *J. Org. Chem.* **18**, 707 (1953).
17. I. M. Kolthoff and W. J. Dale, *J. Amer. Chem. Soc.* **69**, 441 (1947).
18. I. M. Kolthoff and W. E. Harris, *J. Polym. Sci.* **2**, 41 (1947); N. Rabjohn, R. J. Dearborn, W. E. Blackburn, G. E. Inskeep, and H. R. Snyder, *J. Polym. Sci.* **2**, 488 (1947).
19. J. A. Rozmajzl, *Macromol. Syn.* **2**, 57 (1966).
20. L. Horner and J. Junkerman, *Justus Liebigs Ann. Chem.* **591**, 53 (1955).
21. R. J. Coleman, *Macromol. Syn.* **2**, 63 (1966); K. Tessmar, *Kunststoffe* **43**, 496 (1953); *Chem. Abstr.* **48**, 3061 (1954).
22. Y. Ohtsuka, H. Kawaguchi, and Y. Sugi, *J. Appl. Polymer Sci.* **26**, 1637 (1981).
23. H. Kawaguchi, Y. Sugi, and Y. Ohtsuka, *Emulsion Polymers and Emulsion Polymerization, ACS Symposium Series* **165**, 146, D. R. Bassett and A. C. Hamielec, eds. (1981).
24. C. E. Breed, V. S. Frank, and A. J. Urjil, German Patent 1,074,859 (1960).
25. "Vynate Copolymer Latexes For Exterior Architectural Coatings," Technical Bulletin F-60831, Union Carbide Chemicals and Plastics Co., Danbury, Connecticut, 1994.
26. E. H. Riddle, "Monomeric Acrylic Esters," Van Nostrand-Reinhold, Princeton, New Jersey, 1954.
27. "Emulsion Polymerization of Acrylic Monomers," Bulletin CM-104 A/cf, Rohm and Haas Company, Philadelphia, Pennsylvania.

INDEX

ABS resins, 41, 94
Acetaldehyde
  polyacetals from, 151-152
Acetalization, 149, 155
Acrylamide
  anionic polymerization, 91-92
  emulsion polymerization, 88-91
  hydrogen transfer polymerization, 93-94
  polymerization, 72, 81-84
  solution polymerization, 85-87
  suspension polymerization, 84-85
Acrylamide-styrene copolymer
  by emulsion polymerization, 89, 274-275
Acrylic acid
  esters of, 72
  polymerization of, 73
Acrylic acid salts
  polymerization of, 80
Acrylic and methacrylic acids
  bulk polymerization, 74
  solution polymerization, 76
  suspension polymerization, 74
Acrylic esters
  bulk polymerization, 107-109
  emulsion polymerization, 112-116
  group transfer polymerization, 118-119
  ionic polymerization, 116-117
  polymerization, 103ff
  solution polymerization, 111-112
  suspension polymerization, 109-111
Acrylic monomers
  polymerization of, 72-123
Acrylonitrile
  anionic polymerization, 102
  emulsion polymerization, 100-101
  polymerization of, 72, 94-102
  slurry polymerization, 97
  solution polymerization, 98-100
  stereospecific polymerization, 102
Acrylonitrile-butadiene copolymer
  emulsion polymerization of, 274
Alcohols and diols
  polyoxyalkylation of, 207-210
Aldehydes
  polymerization of, 40-44
Alkyd resins, 27, 141-147
Alkylene carbonates
  oxyalkylation reagents, 207, 211-212
Alkylisocyanate polymerization, 35
Allyl acrylate
  polymerization of, 252-253
Allyl esters
  polymerization of, 250-260
  ring formation with, 251-253, 258
Allyl methacrylate
  polymerization of, 104, 252-253
  two-stage polymerization, 250-251
Amino resins, 125-129
Anionic polymerization, 17, 8-12
Aramid fibers, 36-37

Benzaldehyde
  polyacetal from, 151-153
  poly(vinyl acetal) from, 153
Benzyl p-hydroxybiphenylpolyglycol, 55
Bis(3-aminopropyl)ether, 54-55
Bis(chloroethyl) ether
  carcinogenicity, 125, 130
Bis(2-hydroxyethyl) sulfide
  polyacetal with benzaldehyde, 152-153

279

Bisphenol A diglycidyl ether
  curing of, 137–139
Block copolymers, 10
Block and graft polyesters, 29
4-Bromo-2,6-dimethylphenol
  oxidative polymerization, 67
Bulk polymerization
  of olefins, 3–5
Butyl rubber, 16
Buna rubber, 94
Butadiene-styrene copolymer
  by emulsion polymerization, 273–274
Butyl acrylate-vinyl acetate-vinyl neodecanoate
  emulsion terpolymerization, 275–276
p-tert-Butylphenol
  polyoxypropylation of, 213–214
Butyraldehyde
  poly(vinyl acetal) from, 155

Calcium acrylate
  polymerization of, 231
N-Carbamates, 58
Catalysts
  for vinyl ether polymerization, 158
Cationic polymerization
  of olefins, 14–16
Ceiling temperature, 181
Celcon®, 41
Chain-transfer activity
  of t-butyl alcohol, 233
Chain-transfer agents, 106–107
  in styrene latex polymerization, 7
Charge-transfer polymerization, 76–78
Chloral, polymerization of, 41
Coordination catalyst
  in olefin polymerization, 17–19
Cyanoacrylate esters, 95

Dead-end polymerization, 231
Diallyl carbonates
  polymerization, 253–255
Diallyl phthalates
  polymerization, 255–257
Diisocyanates
  polymerization, 54, 194–202
  reaction with diamines, 54–55
N,N-Dimethylacrylamide
  anionic polymerization, 91–92

Dimethyldichlorosilane
  use in silicone resin formation, 178
Diphenyl carbonate, 31

Emulsion polymerization, 266–278
  notes on techniques, 266–267
  recipe development, 268–272
  of styrene, 6–8, 273
Epoxy resins, 135–140
  analysis of, 135
  by polymerization of epoxides, 46–52
Esterification reactions, 25, 58–59
Ethyl acrylate
  emulsion polymerization, 114–116, 276–277
Ethylene glycol
  polyoxypropylation, 209
Ethylene oxide, 46
  polymerization, 49–57
  safety precautions, 203–204

Flame retardants, 60

Glass-transition temperature, 103
Glycerol
  reaction with phthalic anhydride, 27
Glycidyl methacrylate
  copolymerization, 111–112
Guanidines, 198

Heat of polymerization
  of styrene, 3
1,6-Hexamethylene diisocyanate, 54–55
Hexamethylene dithiol
  oxidation to polysulfide, 187–188
Hexamethylolmelamine prepolymer, 128
Hydrogen transfer polymerization, 91–94
Hydroxy compounds
  polyoxyalkylation of, 203–216

Imide formation, 82, 85
Inhibitors
  for acrylic polymerization, 105–106
Initiators,
  anionic, 8–9
  cationic, 14–16
  in emulsion polymerization, 271–272

Interfacial polymerization, 37-38
Isocyanate polymerization, 194-202
Isotactic poly(methyl methacrylate), 116-117
Itaconic acid, 72
  polymerization, 79-80

Kevlar®, 36

"Living" polymerization of propylene oxide, 210
"Living" polymers, 1, 10, 19

MDI, 60 [see also Methylene bis(4-phenyl isocyanate)]
Melamine-aldehyde cocondensations, 127-128
Mercaptans
  oxidation to polysulfides, 187-188
Metallocene catalysts
  in olefin polymerization, 19
Methacrylamide polymerization, 87-88
Methacrylic acid esters, 72
Methacrylonitrile, 94-95
  emulsion polymerization, 103-105
Methyl methacrylate
  group transfer polymerization, 118-119
  suspension polymerization, 110-111
Methylene bis(4-phenyl isocyanate), 60, 62
Methylolmelamines, 127
Michael reaction, 92-93
Microspheres, monodispersed, 12-14

Nitrile rubbers, 94
Nitriles
  in polyamide synthesis, 36
Nomex®, 36
Nonyl phenol polymerization, 213
Noryl®, 65
Novolak resins, 130
Nylon, 34-35
Nylon(1) preparation, 194-198
Nylon(6,10), 38

Octene-sulfur dioxide copolymer, 181-182
Olefin,
  polymerization of, 1
  sulfur dioxide copolymers, 178-184

Organic oxides
  safety precautions, 203-204
Orlon®, 94
Oxiranes, 136
Oxyalkylation, 205

Phenol-aldehyde condensations, 130-134
Phenol-formaldehyde resins, 125-133
Phenols, polyoxyalkylation, 212-214
Phenyltrichlorosilane
  silicone resins from, 177
Phthalic anhydride
  reaction with glycerol, 27
Polyacetals, 148-157
Poly(acrylamide) gels, 86-87
Poly(acrylic acid)
  isotactic, 73
Poly(acrylonitrile)
  monodisperse, 100-101
  solvents for, 99
Polyadipates, 25
Polyamides, 34-39
Poly(arylene sulfides), 188-192
Polycarbonates, 25, 31-33
Poly(2,6-dimethyl-1,4-phenylene ether), 65-68
Polydisulfide, 187-188
Polyesterification reaction, 25
Polyesters, 25-30
Polyethylene adipate (see polyadipates)
Polyethylene oxide, 49-50
Polyethylene terephthalate, 25, 27-29
Polyformals, 149-151
Poly(glyceryl phthalate) alkyd resin, 144-148
Poly(hexamethylenesebacamide), 38
Polyhydroxy compounds
  polyoxyalkylation of, 210-212
Polyisocyanurates, 201
Polyketals, 148-150
Polymerization of isocyanates, 194-202
Poly(methacrylic acid)
  by hydrolysis, 73
  by solution polymerization, 76
  by suspension polymerization, 75
Polyorganosiloxanes (see silicone resins)
Polyoxyalkylation
  of alcohols, 207-210
  of hydroxy compounds, 203-216
  reagents for, 207, 209
Polyoxyalkylene acetates, 46
Poly(4-oxyhexamethyleneurea), 55-56

Polyoxymethylene, 40
Polyoxypropylene glycol, 47–49
Polyphenylene ethers, 65
Poly(m-phenylene isophthalamide), 36
Poly(phenylene oxide), 65
Poly(phenylene sulfide), 188–192
Poly(p-phenylene terephthalamide), 36
Polyspiroacetals, 148–151
Poly(styrene sulfone), 180
Polysulfides (see Sulfide polymers)
Polyureas, 53–57
Polyurethane foam formulation, 60
Polyurethanes, 58–63
Poly(vinyl acetals), 148–157
  alcoholysis, 263–264
  general procedure for preparation, 155
  tacticity, 229–230
Poly(vinyl acetate), 223–227, 262–263
Poly(vinyl alcohol), 217–224, 230–243, 261–265
Poly(vinyl butyral), 148–156
Poly(vinyl ethers), 158–165
Poly(N-vinyl pyrrolidone), 166–175
  complex formation, 173–174
Prepolymers
  of polyurethanes, 60–61
Propylene oxide
  polymerization, 47–49
  safety, 203–204
Propylene-sulfur dioxide copolymer, 182–183

Resoles, 130
Ring opening metathesis
  polymerization, 18
Ritter reaction, 36
Ryton®, 188

Schotten–Baumann reaction, 31
"Short stop," 221
Silicone resins, 176–178
Sorbitol polyoxyalkylation, 211
Stearyl alcohol polyoxypropylation, 209
Styrene polymerization, 1–14
Sucrose polyoxypropylation, 210–211
Sulfide polymers, 185–193
Sulfur dioxide-olefin copolymers, 179–184
Surfactants for emulsion
  polymerization, 270–271
Suspending agents, 5, 110
Suspension polymerization, 266

Syndiotactic poly(methyl methacrylate), 116–117
Syndiotactic polyacrylates, 111

Tebbe reagent, 18
2,4-Toluene diisocyanate (TDI), 60
  polymerization to a carbodiimide, 200–201
Transacetalization, 150

Urea
  reaction with diamines, 55–56
Urea-formaldehyde condensations, 126–127
Urones, 126

Vinyl acetate
  bulk polymerization, 232
  emulsion polymerization, 239–240
  emulsion terpolymerization, 242, 275–276
  initiation, inhibition, and retardation, 222–223
  nonaqueous dispersion polymerization, 243–244
  polymerization of, 217–248
  solution polymerization, 232–234
  suspension polymerization, 235–237
Vinyl benzoate, 223, 227–229
Vinyl n-butyl ether polymerization, 158, 160
Vinyl butyrate, 223
Vinyl esters
  bulk polymerization, 230–232
  emulsion polymerization, 230–232
  nonaqueous dispersion polymerization, 243–244
  polymerization of, 217–248
  reactivity ratios, 227–229
  solution polymerization, 232–234
  suspension polymerization, 234–237
Vinyl ether-maleic anhydride
  copolymer, 162–163
Vinyl ethers
  cationic polymerization, 158–160
  copolymerization, 159, 161–163
  free radical polymerization, 158, 161–163
Vinyl formate, 223
Vinyl palmitate emulsion
  polymerization, 241
Vinyl propionate, 218, 223

N-Vinyl pyrrolidone
  aqueous solution polymerization, 167–169
  bulk polymerization, 167–169
  cationic polymerization, 172–173
N-Vinyl pyrrolidone-methyl methacrylate
  copolymerization, 172–173
N-Vinyl pyrrolidone polymerization initiators, 167–168

Vinyl stearate, 217
Vinyl thioacetate, 229
Vinylidene cyanide
  polymerization and hazards, 95

Ziegler–Natta catalysts, 17–19